QUALITY IS
STILL FREE

OTHER McGRAW-HILL BOOKS BY PHILIP B. CROSBY

QUALITY IS FREE (1979)

THE ART OF GETTING YOUR OWN SWEET WAY (1983)

QUALITY WITHOUT TEARS (1984)

RUNNING THINGS (1986)

THE ETERNALLY SUCCESSFUL ORGANIZATION (1988)

LET'S TALK QUALITY (1989)

LEADING (1990)

PHILIP CROSBY'S REFLECTIONS ON QUALITY (1996)

QUALITY IS STILL FREE

Making Quality Certain in Uncertain Times

Philip B. Crosby

McGraw-Hill
New York   San Francisco   Washington, D.C.   Auckland   Bogotá
Caracas   Lisbon   London   Madrid   Mexico City   Milan
Montreal   New Delhi   San Juan   Singapore
Sydney   Tokyo   Toronto

Library of Congress Cataloging-in-Publication Data

Crosby, Philip B.
    Quality is still free : making quality certain in uncertain times
  / Philip B. Crosby.
        p.   cm.
    Includes index.
    ISBN 0-07-014532-6 (hc)
    1. Quality assurance. I. Title.
  TS156.6.C76 1996
  658.5'62—dc20   95-32400
                                              CIP

McGraw-Hill

A Division of The **McGraw·Hill** Companies

Chapters 1 and 4 first appeared in *Quality Is Free*, copyright 1975 by Philip B. Crosby, published by McGraw-Hill, Inc.

    4 5 6 7 8 9 0 DOC/DOC   0 0

ISBN 0-07-014532-6

The sponsoring editor for this book was Philip Ruppel, the editing supervisor was Jane Palmieri, and the production supervisor was Pamela Pelton. It was set in Fairfield by Ron Painter of McGraw-Hill's Professional Book Group composition unit.

Printed and bound by R. R. Donnelley & Sons Company.

McGraw-Hill books are available at special quantity discounts to use as premiums and sales promotions, or for use in corporate training programs. For more information, please write to the Director of Special Sales, McGraw-Hill, Professional Publishing, Two Penn Plaza, New York, NY 10121-2298. Or contact your local bookstore.

 This book is printed on recycled, acid-free paper containing a minimum of 50% recycled de-inked fiber.

*To my brother David,
the creative one in the family*

CONTENTS

EPILOGUE 227

Guidelines for Browsers   *231*
Index   *259*

PREFACE TO
QUALITY IS FREE

I have learned to carry my typewriter with me as I travel. Renting is iffy, and unreliable at best. Checking the machine with the airlines is not wise. Portable typewriters are just not packaged for "luggage" treatment. Machines that can survive this system cannot be lifted.

Naturally, your traveling companions ask if you are a writer. Now if you really are a writer, all you have to do is say yes and an interesting conversation is established during the trip. However, I have never considered myself a writer, I consider myself a professional manager who communicates through many means; one of these is writing.

That may seem like a small difference in terms, but it is really more than that. Trying to explain your ideas so others can understand them is what this sort of thing is all about. Trying to offer concepts in attractive packages so the communicatee has to at least consider them has been the struggle of my life. Some of these concepts have been accepted, but usually only several years after I began developing them. That is only fair since it takes a lot of years to conceive them.

I wasn't born a manager; my family always envisioned me as a medical person. My father was a podiatrist, my uncle, a physician, and the whole outfit was involved one way or another in the medical field. I grew up assuming I would enter it too.

It is not my intention to relate the story of my life; you would doze off before reaching the end of the page. No life story, but there is a point to all this. I started at the bottom of the business and have had each and every job on the way up. Inspector, tester, assistant foreman, junior engineer, reliability engineer, group engineer, section chief, manager, director, corporate vice president—all of them. This has produced a "dirt under the fingernails" education I would not have received if fate had dealt me relatives who believed in the God of engineering or accounting.

Because of these experiences, I tend to see things in terms of those who must finally wind up doing the job. I see concepts and their implementation as people-oriented. Once in a while I get a glimpse of the future, enough to know what will be accepted and what will be ignored. In preparing this book I have tried to emphasize the practical actions of communicating programs and concepts in a way that will bring results.

This book took an awfully long time to write. Much of the material I put together over several years has been discarded. The Grid is a new development as is Make Certain. Both programs are unique, cheap, and remarkably effective. If you can't communicate with management using the Grid, and with people using Make Certain, well, then you are in trouble too deep to be helped by this little book. Not company trouble, comprehension trouble.

My staff has been very patient with me during the preparation of this material. Virginia Brauneck, my secretary, has struggled through translating my clumsy typing into English. Alternately scowling and breaking into giggles, she put a great deal of herself into this work. I appreciate it.

My leaders encouraged me through their comments and interest. Not all corporations would understand when one of their senior executives sits alone in a hotel room at night making clickity-click noises. My corporation recognizes communication as the engine that makes our society operate or strangle.

More people have helped me through these twenty-five years than I can name here. Three of these individuals have passed on. They were special to me and I would like to publicly remember Tom Willey, Jim Halpin, and Murray Hack.

And, of course, the Crosby family. They always understand; they love me anyway.

I hope you will read the first three chapters in order. It will all make more sense to you that way. After that you can hop around any way you please. After all, it's your book.

Philip B. Crosby
JOHN'S ISLAND, FLORIDA
1978

PREFACE TO
QUALITY IS STILL FREE

Quality became a problem because those who thought and wrote about it considered it to be a separate effort from managing the enterprise. It was made to be very technical, with statistical tools and analytical reasoning. Developed initially for the industrial world, the techniques and concepts slipped into administrative and service life. At the base of all this was a system, called Quality Control, which worked very well in containing the outputs of machine shops and assembly operations. Bad things were found out as early as possible and then reprocessed. However, it was very expensive and inefficient to have all this happen so far along in the process. The basis of this thinking is that variation is in everything and must be measured and contained. Statistics thus becomes the solution as well as the reason. Those who are dedicated to this way of life make it a self-fulfilling prophecy. They waste time grinding away at evaluation instead of thinking about prevention and working on that. This is an inevitability complex hard to stamp out.

Quality Is Free showed that the result of quality lay in the hands of management, not in the Quality Control Department. The book emphasized prevention and cooperation rather than detection and discipline. It made clear that the philosophy and policy management put forth determined what the customer was going to receive. Few believed this until they found themselves with declining market shares and rising customer complaints. At this time they began to realize that what they were delivering was not that wonderful. *Quality Is Free* made that clear, and many appreciated this understanding.

The quality reformation of the 1980s can be attributed to management accepting enthusiastically their role in quality management. Defect levels dropped dramatically; the Price of Nonconformance led to profit improvement. Everyone felt good about the future of quality now that management personally felt responsible for it. But once the fires of complaint were damped and customers were being satisfied with the products and services provided for them, things began to change again. Management began to back away from being personally concerned and returned quality to committees and teams.

Now we find ourselves faced again by systems (ISO 9000 is one; the Baldrige Award criteria is another) that can be installed in a company with the idea that it will take over the job of managing quality. This situation is like in my days in aerospace during the 1950s and 1960s. All companies and their suppliers were certified to Mil-Q-9858A, which the Department of Defense monitored closely. We all had books of procedures to show how we conformed. Yet in all the years I worked in that environment, I have no memory of anyone every using "9858" or its procedures to do anything about running the company. In fact, it always was hard to find a copy to use when the Department of Defense came by for an audit. What we were actually doing was much more than this specification contained. It was not enough, just as the ISO documents are not enough. No system can be installed to make management happen in any function.

We also find that success has brought about a relaxing of the need for Zero Defects in the minds of those who manage quality. The nonconformance situation semiconductor suppliers found recently emerged from embracing the standard of "Six Sigma." This permits 3.4 defects per million components. Why anyone would want to do that is beyond me. But they are now paying the price. When even ordinary chips contain a million or more components, such a standard means that they are all defective. Those who believe that variables have a life of their own are going to have to relent one day and face the real world squarely; products, like life, are what you make them. When management puts forth an attitude of wanting less than perfection, that is what they receive in return.

At the time I was writing *Quality Is Free* (1977–78), I was serving as vice president of the ITT corporation and had been there since 1965. This was a fascinating experience and a great opportunity, one I had never anticipated in the early days. I had been a quality professional and a manager prior to ITT and thought I really understood what quality was all about. But it wasn't until I became an executive, and hung out with others of the same stripe, that I really was able to recognize the bones, blood, and soul of quality. Some areas of the corporation did very well in quality, some were very bad at it. The policies, standards, and supervision were the same for all. The way local leadership thought and acted as individuals determined the quality result of their organizations. Putting in a system to bring about quality was of little use; changing the executive culture was the key to changing the result. It was in this way that my philosophy evolved.

Before all this, in 1952, when I left the Navy and started work on an assembly line, becoming an executive seemed like an impossible goal. I realized early on that the senior level would be the place to get something done but my realistic life work goals at that time topped out at perhaps being a general foreman. However, shortly after entering the business world I realized that the way things worked was not really very practical. The conventional wisdom assumed that nothing would ever be done right, so most of the effort was placed on checking initial results and then correcting them. Industrial life was built around "acceptable quality levels"; service life copied that. It was all done very formally: papers were written about calculated risk, everyone accepted it. This did not compute with the medical background I had accumulated in the past few years.

As a result I concentrated on trying to change the conventional wisdom and in the process obtained results that let me have better jobs at Martin-Marietta in Orlando, and then ITT. These jobs gave me the chance to teach management a better concept of operating. By measuring these actions and results financially (the Price of Nonconformance) it was possible to place quality into the mainstream of a company. That got management's attention. It was to show how this could be done that drove me into writing *Quality Is Free*.

When *Quality Is Free* became a best-seller, I left ITT, with blessings, and started Philip Crosby Associates, Inc. (PCA), in order to help others. PCA became a large public company with many clients worldwide, and we never made a sales call while I was CEO. We dealt only with those who came to us. (The story of PCA and the development of my concepts is covered in Chapters 6 through 11.)

Quality Is Free quickly became a standard. It has been translated into 20 languages and is found in almost every bookstore around the world. I know, because like all other authors, I go look. I have written eight other books between the time it was published and today. Some were on quality but most concerned themselves with general management. I was still trying to explain that quality, meaning doing what you said you would do, is not an added factor but the absolute essential component of a successful enterprise.

When McGraw-Hill approached me with the thought of revisiting *Quality Is Free* it seemed like a fine idea. Upon rereading it again, I found that the concepts and language were still valid but that the case histories and stories could use some updating. I particularly wanted to talk about the "tough company" way of operating that I had been developing.

But my real objective is to revisit the basics and discuss how they have been accepted or abused. Over these past 15 years people have written and talked about the book, and it is credited with beginning the quality revolution. A couple of million copies have been printed and sold. Yet, as other authors have learned, few people actually internalized the content. The title itself, for instance, is still misunderstood, which is easy to do if one thinks that quality means "goodness." "Goodness is free" is obviously not realistic; nor is "delighting the customer is free" or "fit for use is free." Reality is what management is supposed to be about, not opinion. Goodness is not reality.

The dirty little secret of quality management prior to *Quality Is Free* is that quality was not really managed, it was contained. Everyone just played "let's pretend." The quality control, quality assurance, reliability, and other efforts had no effect on anything except to shield the customer a little. They just went on, and still do, sort of like the way drug agencies combat drugs. There is a lot of smoke and fire, but few solid results. More bad stuff came in and went out every day. The only solution to quality is to work diligently at learning how to do things properly while preventing problems in order to meet the customer's need. The only solution to the drug problem is to eliminate the customer's need. Doing this has nothing to do with shooting down airplanes, busting doors, and locking up dealers. Quality management has nothing to do with inspection, material review, system certifications, and other recreational activities. I am going to reposition *Quality Is Free* for the nineties and beyond. Perhaps this time I can make it so clear that it cannot be misunderstood.

I would hope that the reader would remember that all the things I write about have emerged from personal experience. It has been proven through the centuries that we have a better chance of learning about real life if we have a teacher who actually lived real life. To place one's faith and career in concepts and guidelines that have never had to actually work is not a responsible plan. Ask yourself how they learned what they teach—where did it come from? Can someone who has never been married offer worthwhile marital counseling?

I would like to thank my wife Peggy for her patience and support; Debbie Eifert, my assistant, for her consistency; and the McGraw-Hill team of Publisher Philip Ruppel and Editing Supervisor Jane Palmieri for their guidance and help.

Philip B. Crosby
PhilCros@aol.com
HIGHLANDS, NORTH CAROLINA
WINTER PARK, FLORIDA
1995

QUALITY IS
STILL FREE

REVISITING "MAKING QUALITY CERTAIN"

The phrase "Making Quality Certain" means creating an organization where transactions are completed correctly the first time; where employees, suppliers, and customers are helped to become successful; and with whom everyone is proud to be associated. None of these characteristics were part of conventional concerns of management at the time I wrote *Quality Is Free*. Quality was something that happened or didn't happen, the laws of probability were against it anyway; customers got what they were considered to deserve; change came about reluctantly, if at all; and very few employees, suppliers, or customers liked any company. In managerial minds everything was supposed to be a struggle; people were considered to be generally unreliable and expendable. None of this mattered since there was an unlimited supply of them. As part of this mindset, whatever one department received from another department inside a company was normally incomplete, late, and often inaccurate. That is the way life was supposed to be.

I didn't like the continual hassle of business the way it was run, the resourceful snatching of victory from the jaws of defeat day after day. I liked an even flow, where everyone knew what they were doing and then did it. Those who had been in business longer than I tolerated my childish beliefs; their experience had taught them about what they considered to be the real world. It was not a great place to have an even and purposeful life.

But I had an insight, and sometimes that is better than experience. And my experience, after that insight was revealed to me, showed that it was correct, or could be made to be correct. The insight revealed that all of this activity was contrived because everyone thought there was no way to get things right. To me this was just plain wrong. Individuals really wanted to do a proper job and, if given the proper goals, information, and leadership, they would do just that. Management assumed that it was easier to spend energy and resources fighting situations than it was to take actions that would prevent problems. They also believed, and defended the concept fiercely, that people were genetically incapable of defect-free performance. They felt that people had to be overcome by some established way of doing things, called systems. They put their faith in these systems rather than in people.

Those who taught how to manage, those who managed, and those who conducted professional activities within organizations had this in common: they believed that the determination along with application of the correct system would make everything come out as planned. And, of course, the goal of all this was to come close to getting everything right. They were forever writing procedures, conducting classes, creating software, and having programs. When results did not match what had been anticipated, the system was tuned up some more, the people were "revitalized," and some sparkling new technology or code words were obtained. When all else fails, management finds a new, and purported to be advanced, system, perhaps leaping on the latest set of initials to come out in book form. This constant search for the Holy Grail is what turns off the "middle" management. They find that they are always busy working on the previous fad; senior management discards those systems the way debutantes dismiss swains. After a few of these experiences they learn to be wary and not get involved until the boss proves intent.

The reality of business life is that systems fail because business life, like personal life, is chaotic—it doesn't react to systems, it does what it wants to do. It is like the guy who studied oriental self-defense for several years and became an expert at it. However, when braced by a mugger in the park he dropped to his knees and screamed "don't hit me." When it came time to implement the system he reverted back to chaos. That is entirely normal. A visit to any office, factory, pretzel stand, band, military group, or other organization reveals that everyone is struggling to make things happen. What comes into the place is never exactly what is supposed to come in. Things are late, early, or wrong, and the work atmosphere is uncooperative. People must learn to react to what is going on; they must have a base from which to function. They do not have time to remember and apply the systems. Those who have been in armed combat can easily recognize that the enemy does not always follow the plan. We have to be prepared to make the right move when they make the wrong move. That way we don't fight our people, we fight the chaos of the world.

My original purpose in writing the chapter "Making Quality Certain" from *Quality Is Free* was to show that quality does not come about as the result of some particular way of dancing. People have to be helped to understand that they reside in a culture where the correct action and result are desired. "Making Quality Certain" means that the basics have to become part of the woodwork. The way to prevent fire fighting is to not have fires. No professional fire-fighting group has a strategy of causing fires and then taking credit for putting them out. Given the slightest opportunity to get out of their rut, people will launch into an attack mode rather than spend their life on the defensive. It is much more rewarding, as well as fun.

The idea of causing quality to become a normal part of an organization's operating arsenal did not catch on automatically. Just reading about it was not enough, management had to come to a clear comprehension. For this reason I founded Philip Crosby Associates (PCA) in 1979 and created the Quality College as part of it. Executives and managers would come to class for several days where we would have time to help them understand that their ideas about quality were incorrect. It wasn't that they just had to find a system and keep it tuned, it was that there is no system to make it happen. Going through the Baldrige criteria, the

ISO series, or some similar list of activities results only in books of procedures and wasted efforts in urging people to comply with them. That is quality control, not quality management. When someone asks me about getting "certified" or "approved" to such things, I always ask them what they are going to do with it. All the automobile drivers in the world are "Certified," which has nothing to do with their capability of driving or their system for doing it.

Rather than coming from systems, quality actually emerges from the way management presents policy, education, and their personal example. It is a reflection of the standards of the management, up or down. The quality professionals, for the most part, and those who teach them, didn't understand or agree with this. They like specific activities and official programs such as Quality Circles, Empowerment, Team Building, Benchmarking, and such. Then if things don't get better, they can look for another set of initials. They adore "approved" systems such as ISO 9000 and Mil-Q-9858 which can take the blame for lack of results. That is why they become addicted to "gurus" who have systems.

Mil-Q-9858, birthed by the U.S. Department of Defense (DOD), has been around for 40 years and there is not a single case I know of where it caused a company to produce conforming material and services routinely. In fact, it moves operations in the other direction because it forgives nonconformance and provides for formal ways of accepting things that are not correctly accomplished. This success pattern led the DOD to create what they call TQM which is following the same pattern of activities, procedures, forgiveness, and confusion. I refer to it as Trivializing Quality Management.

Quality had always been conducted as though it were a difficult task in organizations, whether they considered their business to be service or manufacturing. The subject was viewed as heavily technical, loaded with statistical analysis, and considered to be beyond the grasp of anyone but seasoned quality-control professionals. They spoke a language all their own and were primarily oriented toward the inevitability of error. It was my hope to change all that in *Quality Is Free* by showing that putting the emphasis on preventing problems, rather than finding and fixing them, was where progress lay.

Prevention, as a work ethic and practice, must be deliberately inserted into the operating culture of an organization if it is to have an effect. I understood this from having had the responsibility for achieving quality in several companies over a 25-year period. Traditionally this was a cops and robbers assignment. I changed that by treating the operating departments as though they were honest people trying to do the right things. They all reacted with strong support and the task of bringing about quality became a pleasure. However, this positive attitude was not shared by many of my colleagues. They went ballistic at the thought of getting along with the enemy. They wanted bigger prisons, more authority, and a death penalty. A "007" license would be considered appropriate. Then when some department head refused to obey the quality-control rules, he or she could be quietly exterminated which would serve as a lesson for others.

This divergence of concern between the conventional wisdom of quality and myself began when I proposed the performance standard of Zero Defects back in 1961. I, along with the concept itself, was attacked by the quality-control profession's brightest lights. It was as though I was trying to reverse the flow of the Mississippi River. Acceptable quality levels were formalized and inevitable. The "trade-off" between good and bad was considered a part of doing business. Everyone had rework stations by one name or another—chemical companies "blended," for instance. Waste was routine. Automobile companies were spending at least $2000 a car just on repair during assembly, plus warranty and recall expenses after the sale. Service companies prided themselves on their customer service operation, which was rework all the way. Senior executives did not recognize nor believe that their concept of operating was causing the problem. They thought they just needed better workers. As a result, management and workers distrusted each other. Changing this ingrained belief was my goal. Simply, we just had to learn how to get the right things done right, the first time.

To me it was only common sense; others did not see it that way. They felt mine was a simplistic approach, but this was because no one had ever written a book on quality without charts in it before. I didn't even offer a fail-proof system. When executives read *Quality Is Free* after it came out in 1979 I received many calls asking for help in making quality certain.

When we had people at the Quality College for a few days they would begin to understand the idea that quality cannot be accomplished by some separate campaign. Thousands of managers have gone back and successfully built quality into their operations. Hundreds of companies have profited from their efforts. But this comprehension came because we spent time overturning concepts that were buried deep inside them. It is not an easy task; people are not willing to change by just reading something that sounds sensible. When General MacArthur's group asked Homer Sarasohn to come to Japan and teach their top management about quality, he kept them locked up for weeks until they got the old ideas out of their minds. It was making these executives understand that all work is a process that turned Japan around. Dr. Ishakawa of the Joint Union of Scientists and Engineers led the quality revolution from that point. Others have leapt forward to take credit, but these men were the real leaders.

Today most company management are still struggling with quality, expending a lot of effort on systems and activities while achieving small results. My estimate is that they get 25 percent of the return they could achieve if they worked on the right things. However they get off track because of illusions.

Management has the illusion that quality can be accomplished by a system such as TQM, or code words such as "continuous improvement." They think that if they appoint a committee to handle quality, it will get handled. They would not dream of doing this with finance.

Implementors have the illusion that quality is caused by activities and programs such as Benchmarking, Empowerment, Statistical Process Control, Team Building, and such. They mistake activities for change and as a result cause very little change.

Practitioners are continually coming up to me to relate that they think my concepts are realistic and correct. They state that they are completely in agreement with me. However, a few casual questions usually reveal that they are not doing that in practice. It is hard to be different in public.

Governments, and associations, have the illusion that they can "assure quality" by imposing a specification such as ISO 9000, or an award program such as Baldrige, which is supposed to contain all the information and actions necessary to produce quality. This is the ultimate in naiveté. No one would accept something similar for finance, marketing, administration, or any other function. However, getting companies qualified to such specifications or awards does provide a nice living for those who consult in the field.

Prevention comes from a lifestyle, which evolves from management principles and guidance, which come from an understanding of what it takes to make the business profitable and the company tough.

The explanations and stories contained in the original first chapter of *Quality Is Free* are valid and applicable. For that reason we have reprinted Chapter 1, "Making Quality Certain," from *Quality Is Free* in its entirety. Any references in the following chapter to the book's content are references to content in *Quality Is Free*.

MAKING QUALITY CERTAIN

What does "making quality certain" mean? "Getting people to do better all the worthwhile things they ought to be doing anyway" is not a bad definition. "People" includes top management as well as the lower levels of the organization. After all, part of the top job is making certain that all management functions have the opportunity to perform their responsibilities. The problem is, of course, that everyone who arrives at a top management job gets there by moving up through one division, like finance or engineering, that has a limited, specific function, and may or may not have any ideas about overall quality. Top managers may or may not realize what has to be done to achieve quality. Or worse, they may feel, mistakenly, that they do understand what has to be done. Those types can cause the most harm.

It is up to the professional quality manager to assume the responsibility for instructing top management about this portion of their job. It is not necessary to be extremely clever or brave to accomplish this; it is only necessary to be able to explain it all in terms that cannot be misunderstood. Professionals in any role who obscure explanations by using mysterious terminology do themselves, and their roles, a disservice. They get some satisfaction from seeing obvious confusion on the face of their superiors, but that confusion just makes everybody's job harder.

I started in the quality business as a junior technician testing fire control systems for B-47s. Completely untrained and uninformed, I learned the simple tasks of adjustment and measurement without ever really wondering why it was all being done at all.

In fact, during my first four or five years in jobs like that, it never occurred to me to wonder. But then I had the opportunity to be exposed to reliability concepts and practices. Most of them were rather mushy and mathematical, but they revealed an element I hadn't thought about before: prevention.

That thought brought out a possibility I had never dreamed existed: "Why spend all this time finding and fixing and fighting when you could prevent the incident in the first place?"

The entire world, it seemed, was convinced that prevention—at least on a grand scale—was highly desirable but completely unattainable and impractical. It was always referred to as a sort of dream along the line of King Solomon's lost diamond mines. I had a great many long and earnest talks with sincere people who were clear that there was no way to attain true quality through prevention: "The engineers won't cooperate." "The salesmen are untrainable as well as a little shifty." "Top management cannot be reached with such concepts." "The quality professionals themselves do not believe it."

I knew immediately that I had found the opportunity I was looking for. Here was a problem that everyone wanted to solve but felt was not their responsibility. All I had to do was find a way to get them committed to improvement without having to reveal that they had been wrong all along.

For the next several years, as I learned more about managing quality, I realized that the conventional approach was not effective. Quality managers proudly stood up and announced that they personally were responsible for quality in a particular operation. Just as regularly, and not so proudly, they were sent down in flames when they were unable to resolve all the "quality problems" of the company.

As a project quality manager, I was berated each week by the program director in his staff meeting for not meeting desired goals while the real culprits from engineering, manufacturing, and sales hid their yawns and wished the whole thing would go away so they could return to their important work.

It was all too clear that some beliefs are so basically ingrained that they cannot be changed just by suggesting they are wrong. (I should note that my knowledge of this fact is part of the reason I have been very supportive of the activities of minorities and women in trying to throw off the roles assigned and attributed to them.) However, my active revolution as quality manager didn't really begin until the day one of the company lawyers told me, in all sincerity, that he couldn't really understand "what a bright guy like you is doing in a little cul-de-sac like quality." If I had ever thought of leaving the quality business, that killed it. Some changes had to be made.

So I began to concentrate on the real problems. First, it was necessary to get top management, and therefore lower management, to consider quality a leading part of the operation, a part equal in importance to every other part. Second, I had to find a way to explain what quality was all about so that anyone could understand it and enthusiastically support it. And third, I had to get myself in a position where I had a platform to take on the world in behalf of quality.

I think all these goals have been attained. As a member of senior management of one of the largest industrial companies in the world, I make as much money, and have just as many rights, as other senior managers. We have installed effective and routine ways of understanding quality, and communicating from the top of the organization to the bottom, as well as the other way around. I have not been accused, in the past five years anyway, of having some "quality problem" that I should do something about.

You can do it too. All you have to do is take the time to understand the concepts, teach them to others, and keep the pressure on for prevention. It helps if you train yourself to be articulate, and it helps if you can keep from becoming emotionally involved in the problems of others. But the whole of it is attainable and highly practical.

This book is structured to lead you directly through all the actions required for a proper quality management program. Case histories, all based on my personal experience, explain practically everything so you can see how others reacted in real situations. One of the most interesting of those case histories involved installing a quality management program in the ITT Corporation. I include it here without listing any names of those involved because there were just too many. ITT, at this writing, employs 350,000 people and has yearly sales of over $15 billion. There are 2500 or so executives and over 200 senior executives. You will have to take my word for it that everyone participated. If I listed all their names, it would look like the San Francisco telephone book.

I will tell the story primarily to give background evidence supporting the basic premise of this book. Quality is an achievable, measurable, profitable entity that can be installed once you have commitment and understanding, and are prepared for hard work. The case history is a record of strategy and effort, not a personal job résumé.

In 1965, the top management of ITT decided that they wanted to do something about quality on a corporate basis. It was apparent that quality was a missing ingredient in the corporate sense of things that were important. It wasn't that quality was deliberately considered unimportant; no one was against it. But as an ingredient of industry, like labor, manufacturing, engineering and so forth, it didn't exist. To me, however, quality is the all-important catalyst that makes the difference between success and failure, and my first goal was to create a corporatewide concern for quality. This meant that absolutely correct requirements would be established and would be absolutely conformed to, and that everyone would want to do things right the first time. This concern had to become a part of daily life.

Four objectives were established for the ITT quality program. These objectives have served well through the years, and I commend them to other objective makers:

1. Establish a competent quality management program in every operation, both manufacturing and service.

2. Eliminate surprise nonconformance problems.

3. Reduce the cost of quality.

4. Make ITT the standard for quality—worldwide.

These objectives could not be accomplished by assembling a huge staff at headquarters for the purpose of strangling every potential problem in its crib. There was only me and the secretary I shared with two other guys. It was sort of like assembling a raft from the material you obtained while being swept down the rapids.

So I embarked on a deliberate strategy of establishing a cultural revolution—a cultural revolution that would last forever and become part of the corporate woodwork. Fire fighting would have to be replaced with defect prevention; quality would have to be recognized as a genuine "first among equals"; the habit of doing things right the first time had to become routine; and, most important of all, the whole thing had to happen within the units (ITT's word for subsidiary or other companies) because they wanted it to happen.

To me, a complete corporate quality program has always been a "table" containing all the "integrity" systems. Quality control, reliability, quality engineering, supplier quality, inspection, product qualification, training, testing, consumer affairs, quality improvement, and metrology; and all the other systems and concepts of quality rest on this table. Management selects what it needs from each and applies these tools to its total problem. It isn't necessary or wise for each and every operation to have exactly the same quality program. At ITT, for example, the personality and needs of one unit may bear little actual work relationship to those of another, yet they all need programs both appropriate to them and effective in terms of the total corporation.

To establish such a program requires much more knowledge and participation than just a listing of the tools available in our workchest. It requires that this *integrity systems* table be supported by four pillars, or legs, and that they all be constructed to complement each other. Although these were actually built as part of the same operation, we will discuss them one at a time. The four legs are:

- Management participation and attitude
- Professional quality management
- Original programs
- Recognition

MANAGEMENT PARTICIPATION

"Participation," rather than "support," is the right word for this leg. Management has to get right in there and be active when it comes to quality. Those of us who work for others are liable to monitor and measure them constantly. We examine them continually to determine which attitudes and beliefs are the stronger. We want to know what pleases them, or, perhaps more accurately, what displeases them. And we get very good at finding and calibrating this information. Therefore, causing management at all levels to have the right attitude about quality, and the right understanding, is not just vital—it is everything.

The first struggle, and it is never over, is to overcome the "conventional wisdom" regarding quality. In some mysterious way each new manager becomes imbued with this conventional wisdom. It says that quality means goodness; that it is unmeasurable; that error is inevitable; and that people just don't give a damn about doing good work. No matter what company they work for, or where they went to school, or where they were raised—they all believe something erroneous like this. But in real life, quality is something quite different. Quality is conformance to requirements; it is precisely measurable; error is not required to fulfill the laws of nature; and people work just as hard now as they ever did. These concepts are covered in detail in the following chapters. What should be obvious from the outset is that people perform to the standards of their leaders.

If management thinks people don't care, then people won't care.

At ITT, most of our actions during the formative years were directed toward dispelling the erroneous beliefs and replacing them with those capable of supporting the integrity systems table. We conducted seminars throughout the ITT world on a regular basis. Those managing directors and general managers who had participated in programs and who had learned to understand quality properly testified to the others. They became involved in this evangelic crusade and the word spread: "The programs actually do work and you can trust the quality guy." In 1967, one other quality executive joined the staff and in 1968, quality was set up as a corporate department of its own. At that time, three senior quality managers were brought in from the units to become part of the operation.

Working on a group-by-group, unit-by-unit basis, we work our way through the corporation. Orienting, helping, talking, guiding, badgering, and whatever, we kept the pressure on. New managers joining the corporation were made to feel that participation in the quality program was routine and expected. Therefore, they just sailed right along. Today you would have difficulty finding anyone at the executive level anywhere who hasn't been exposed to the true belief.

PROFESSIONAL QUALITY MANAGEMENT

In the early days, it was not possible to find many of the quality people in ITT units, since they were buried inside the technical and manufacturing operations, if they indeed existed at all. When found, most of them were not permitted to travel. And so we formed quality councils on an area basis. Both in the United States and in Europe, quality professionals joined together to help each other and to determine the types of programs required from the corporate staff. Today there are twenty-seven councils set up by product line or service, and some councils grouped by country also still operate. In addition, there is an executive council on each continent consisting of the chairpersons from all the quality councils. Communication among all branches is comfortable and positive.

To support the councils, and the programs, we instituted the ITT Quality College. Teaching such courses as Quality Management and Product Qualification, the college has issued certificates to over 24,000 people in its lifetime. It is the backbone of the total effort. Every time we think there is no one left to instruct, we find the enrollment full again. The program has been very effective. All the quality professionals of ITT understand the programs the same way. They have been freed organizationally, and report at least on the same level as those they are measuring. This increases the possibility that the programs will be implemented properly.

ORIGINAL PROGRAMS

Traditional quality-control programs are negative and narrow, and it was no different at ITT. Primarily oriented toward product performance, they often turned off the management they were supposed to entice. To overcome this, we constructed numerous programs involving practical activities which could be implemented at the unit level.

"Quality improvement through defect prevention," a 14-step program for improvement, is the foundation of all ITT quality programs. Described in detail later in the book, this program has been implemented in every industry of business of the corporation. Some have been very successful; some have not done as well. But none ever got worse.

It takes four or five years to get people to understand the need for, and learn to have confidence in, such an improvement program. I originally sent a 16-page brochure accompanied by a tape explaining the zero defects concept to every unit. The results were amazing. No one paid the slightest bit of attention. None of them were even sent back. It was apparent that the conversion and instruction had to be done on a unit-by-unit basis until we could get some success stories to use in the seminars.

Other original programs developed were Buck a Day (BAD), a cost-reduction idea program; Zero Defects-30 (a 30-day program contained in one box with enough material for a supervisor and eight to ten of the supervisor's people); Consumer Affairs; Environmental Quality Self-Audit; Quality Management Maturity Grid; Model Quality (a system for printed circuit board manufacturing); Service Company Quality Improvement; Make Certain; and many others.

RECOGNITION

This vitally necessary component of any quality program is often overlooked or conducted improperly. Done correctly, it becomes the shining star of the entire integrity system. We established the Ring of Quality program in 1971. The initial thought was to give recognition to those people who had offered outstanding support to the quality program for a period of five years, or had accomplished one sensational, specific, and unique act. However, it quickly became a program where the winners were nominated by their peers. On that basis, we have processed thousands of nominations and have awarded 182 gold rings to winners. We have also presented several hundred silver pins and citations to other nominees. In every case, we tossed out those who were nominated by their subordinates. Peer nominations make it all come out right.

The Ring of Quality program is taken very seriously. The awards are presented at a formal dinner by the corporate president or chairman. For many of the recipients, it is literally the biggest moment of their life. Like the rest of the program, the presentations are treated with dignity and respect. Reaction to the rings and pins has made one thing very clear: Cash or financial awards are not personal enough to provide effective recognition.

Developing and implementing the four legs of the table

involved traveling millions of miles, talking thousands of hours, and eating tons of food. It was well worth the trouble, as the following comparison of results obtained with the initial objectives should demonstrate.

ESTABLISH A COMPETENT QUALITY MANAGEMENT PROGRAM IN EVERY OPERATION, BOTH MANUFACTURING AND SERVICE. When we began, about 5 percent of our companies had quality programs that could be considered acceptable. In 1977, better than 85 percent were in that category. Pioneer programs were established for the first time in hotels, insurance, car rental, and other service industries. The number of competent quality management and professional people has grown to where their availability is no longer a problem.

ELIMINATE SURPRISE NONCONFORMANCE PROBLEMS. Surprise nonconformance problems have disappeared. We still have problems, and some of them are pips. But never is one all grown up before we find it.

REDUCE THE COST OF QUALITY. The cost of quality (the price of nonconformance) is the expense of doing things wrong. It is the scrap, rework, service after service, warranty, inspection, tests, and similar activities made necessary by nonconformance problems. Between 1967 and 1977, the manufacturing cost of quality at ITT has been reduced by an amount equivalent to 5 percent of sales. That is a great deal of money. The savings projected by the comptroller were $30 million in 1968; $157 million in 1971; $328 million in 1973; and in 1976—$530 million! We had eliminated—through defect prevention—costs amounting to those dollar figures.

Now obviously not all of this was accomplished just by the quality people in the units. Rework people disappeared when there was no more rework. Warranty costs stopped when the properly qualified products didn't fail in the field. There were and continue to be many contributions.

But the facts of life today are that each year your cost of sales rises faster than your prices. That means you have to eliminate or reduce costs in order to make a profit. The best single way to do that is by defect prevention.

Results like these are why I say that quality is free. And not only free but a substantial contributor to profit.

MAKE ITT THE STANDARD FOR QUALITY—WORLDWIDE. The last original objective was to make ITT the standard for quality— worldwide. In trying to determine how to show how we stand, it is fair to ask: "Who says so?" Obviously, we are far from completely achieving this goal, but there has been a lot of progress.

In Europe, all customers for telecommunication are government administrations. In 1965, they were inspecting everything we did. They had resident quality people in every plant we had in Europe, in every country. Today, they let us do the final inspection and testing everywhere. In many countries, they have actually issued us the inspection stamps, and just audit once in a while. These administrations tell our competitors that their operations should be as good as ITT's.

- The Russian Ministry of Electronics surveyed Western quality management systems, and then came to ITT to ask that we come show them how to do it.

- When McGraw-Hill was producting a new encyclopedia of professional management, they asked ITT to write the chapter on quality management.

- American Express said five years ago that Sheraton was the worst hotel chain when it came to quality. This year they rate Sheraton as the best.

- Other companies continually ask for information. During a typical year we receive over 400 requests in the corporate office.

All these achievements represent the result of a great deal of planning and plain, old-fashioned hard work. But it wasn't just planning and hard work that made it happen. One of the most vital components was our success in constructing the first leg of the integrity table—top management commitment.

One of the reasons I cheerfully share these programs with other companies is that I know that many will probably not be able to use them. Not because they are not capable, but because they do not have a top management willing to be patient while the program is ground out four yards at a time. It took five to seven years of unrelenting effort to achieve the cultural revolution at ITT—and I seriously doubt if it will ever be eliminated there.

We recognize that our top management is special because once they understood the realities of quality, they supported the projects, participated in them, and encouraged us all the way along.

The details of how, why, and what are contained in the following chapters. This very brief overview of the strategy behind the ITT program was put in only to show a little bit about how the program all fits together. I know that if I had had the Quality Management Maturity Grid several years ago, the job would have been finished earlier. I know if I had had the ITT experience to play "show and tell" with, it would have been less of a problem to obtain attention.

You have them. Take advantage of what has gone before. Why not learn from the past?

REVISITING "QUALITY MAY NOT BE WHAT YOU THINK IT IS"

Business consists of transactions and relationships, as noted. Quality management's purpose is to cause all transactions to be complete and correct, while all relationships are to be successful. If we understand those two sentences we know all we need to know about quality management. This comes under the heading of something else that should be easy to understand: you can lead a horse to water but you can't make him drink.

The original chapter from *Quality Is Free* made the point early on that the problem was not what people knew about quality management, it was what they thought they knew about it. The conventional attitude of the day was that the workers were sloppy and didn't care; that customers were unreasonable; that closer checking would find problems earlier; and that there was an economics of quality, which meant that you could only afford to get so good. Today, after improving for a while, the situation has deteriorated back to that level of thinking. Quality is an irresistible subject for people at all levels of experience. It has created an entire industry in media alone. New magazines emerge regularly; books appear each month; traditional business magazines have cover stories reinforcing these erroneous assumptions and giving them credibility. Some writers become expert enough to write books on the subject, all without ever having lived a day in a quality professional's shoes. They do it secondhand so long that they begin to think they understand it. At that point they never will shed these wrong ideas. Each article on quality picks up the mythology that has gone before. They rarely see any difference, for instance, in quality control, quality assurance, reliability, and quality management. However, those who ran organizations for a living did.

People believe what they believe, but most of it is not true. A recent study in California on pregnant, unwed teenage girls showed that the fathers of those babies were adults in over 70 percent of the cases. The national view of teenage boys impregnating teenage girls is just not true. If we are going to stop this trend we must deal with the real culprits, not those who appear to be. Stomach ulcers have long been treated as something caused by stress and bad eating habits. When two doctors realized that they were caused by bacteria and cured ulcers by a two-week process of antibiotics they were ridiculed by their peers. Yet the establishment has been wrong all these years, passing information on ulcer treatment along like kids do with sex data. This patter repeats over and over, always because people do not come equipped with open minds.

When Philip Crosby Associates (PCA) first began we received a call from the CEO of a company that made baked products in large volume. She said that her biggest customer was unhappy. The pies were inconsistent in their texture and fruit content. Also, deliveries were not always on schedule, and billings often had mistakes. She had just taken over the company's direction from her father who decided to retire. The customer's representative came to see her right at that time. In addition to all that, she was pregnant.

On the way home from work her first day as the big boss, she stopped in a bookstore and, while searching for something to help, ran across *Quality Is Free*. After reading a few chapters, she called PCA in Winter Park to ask if we could help her. We set up a schedule for her to come over and chat, which she did that week. It became apparent very quickly that her management team's concept of quality was the primary problem. They had internalized each and every one of the erroneous assumptions listed in Chapter 2 of *Quality Is Free* and had been running the company for years based on the inevitability of having conformance and execution problems. When they came for a one-week quality management class at the Quality College, they were kicking and screaming all the way. They sat in class with arms folded for the first two days and then began to get the idea. (That was the kind of reaction we saw in most students over the years.) The CEO got some coaching on the side and attended class with her people. In the meantime, we trained a few of her people to facilitate the course for employees. She drove the whole effort by her obvious intensity.

The customer agreed to hold off for a little while to see if this quality push would bear fruit, at least in the pies. When the managers returned, the CEO made clear once more that they were going to learn how to produce products and professional services that were exactly what the customer had been promised. The managers started by asking for cooperation from the workers to produce pies exactly like the requirements. At this time they quickly learned that there were no requirements. Pies were assembled and produced according to customs that had sort of evolved over the years. This was not only true in the factory, they discovered that the same pattern existed in the accounts payable and receivable departments.

So the quality push began with management having to sit down and describe what a pie looked like and how it was made. They did the same for all the rest of the product line. With the workers' happy assistance, management set the process up, taught everyone their job, and helped it all come off. Within a few weeks everything was better; in a few months it was near perfect. The customer was so happy they started bringing other suppliers (of services and products that did not compete) through that organization to show them what could be done. These people were amazed to find that there were no complex statistical charts, no rework areas, no punishment, no formal team meetings, no expensive quality program. All they saw were management and employees working together to do what they were supposed to do, with the CEO leading the way. When expansion began, each new employee was carefully oriented to this way of working. The CEO's biggest problem became handling all the speech requests that poured in on her.

Let's look at how these original "erroneous assumptions" have fared since 1979. Too bad there aren't 10 of them—I could do a David Letterman–style "top 10 reasons why quality management is not understood." However, here are the five top reasons.

1. *The Belief That Quality Means Goodness, or Luxury, or Shininess, or Weight.* Lack of an agreed definition has been the biggest problem in accomplishing quality management. It is the only area of management that has this problem; most books on the subject never even mention a definition. Those that do, use a full paragraph or two to lay it out. This deeply embedded lack is still there and has become even more embedded. Currently, definitions such as "delighting the customer" and "satisfaction" are tossed about. The fact that these definitions are not measurable or expandable, and are impossible to communicate, has not changed many minds. Telling the designers that the goal is to "delight the customer" is perfectly acceptable; they can do research and work toward that goal. Once they have designed what delights the customer they need to insist that everyone else conform to the requirements so the customer gets what was promised.

The companies that have had successful culture transformations realize that quality means "conformance to requirements." They recognize that requirements are the details of the business that result in customers and coworkers receiving what they have been led to expect. They recognize that management has the responsibility to cause clear requirements to be created, taught, and continually improved. That's what this phrase (which was in the original book) means. Many take it to refer to improving the performance standard a little at a time, as in dropping one baby less each day.

Many writers have taken my definition of conformance to requirements, and twisted it as "conformance to specifications."

This creates a narrow, highly technical, manufacturing-oriented consideration. After misquoting me, they go on to say how stupid it is. Now the problem here is that instead of questioning me in order to find out what I really mean if they don't understand it, they just assume I'm wrong and take off in another direction. This assures the continual discussions of "good," "bad," and "levels" of quality. It lets people come up with silly definitions like "the customer says what quality is." How can you manage with that one? An inexact definition turns quality management processes into events instead of sound management discipline.

The whole purpose of doing quality management is to build a custom and ingrained habit among the employees and suppliers of doing what they said they would do, which is conforming to agreed requirements. Then as we learn how to define better products, processes, and services for our customers and ourselves, we routinely accomplish them. Also the organization is able to deal with the chaos of business life since it can count on things happening as planned.

2. *The Belief That Quality Is Intangible and, Therefore, Not Measurable.* When I conduct a seminar, or talk to people in an airport, or see them in a hallway, the most common comment I receive concerns their management's lack of participation in the quality process. I always ask them if they have included the "price of nonconformance" (PONC) in the financial measurement of the company. They always answer no. Management is not interested in anything that does not have dollar signs on it. Their personal performance is measured by money; they measure everything else that way. Yet very few companies have picked this up.

There was a recent cover story (August 1994) in *Business Week* magazine talking about a "Return on Quality." This had to do with what you get back from doing quality management, and it all made a lot of sense. I did notice that they used a lot of my material without mentioning from whence it came; writers are sensitive about things like that. I choose to take it as a compliment that my thoughts have found their way into common usage. However, the actual measurement of what it costs to do things wrong and over is still not in common usage in companies, and this is a big mistake. My estimate that companies spend one of every four dollars on this item has been proven to be low over the years. Each PCA client was required to learn how to measure quality financially. Service and administrative organizations show 40 percent and more of the operating budget going to waste; manufacturing is an easy 25 percent of revenue. When the comptroller shows all the numbers during the monthly management meeting, PONC should lead the way. That keeps them focused.

But the main reason for determining PONC is to know where to concentrate prevention efforts. I had a computer company as a client and their proudest boast was that their field service people could fix anything. A third of their employee compensation was spent on these folks. We pointed out that this was all rework and showed that the reason units required fixing was that they were not completely finished when they left the plants. Customers did not appreciate having to put up with this.

Similarly, a hotel had a hotline where room problems could be brought to the attention of an assistant manager, who would immediately correct the situation. Learning how to prepare rooms properly the first time resolved that expense. But until they learned what it cost to do it over, they thought they were running things properly. In 1984 I wrote *Quality Without Tears* to show the way to determine PONC easily.

3. *The Belief That There Is an "Economics of Quality."* The conventional wisdom of quality over the years held that one couldn't afford to get too good. This was based on the concept that quality came from appraisal. Those who taught quality control were founded in statistics and in that field nothing is ever 100 percent finished. So management believed in "trade-offs." They would shake their heads at me and emphatically let me know that it would "cost a fortune" to do everything right. All this is pure foolishness designed to eliminate the need to understand the processes involved in their business. It also lets them cut back on training with a clear conscience.

It was the frustration created by this belief that led me to create the Zero Defects concept in 1961. Suppliers would come to me, as quality manager of the weapons system project, and ask for their "AQL." This customary acceptable quality-level calculation stated what percentage of the product or service they supplied to us was permitted to not conform to the requirements. Usually this was expressed as 1 percent, 1.5 percent, 2.0 percent, and such. The original idea of an AQL was to permit things to be sample-checked rather than being examined 100 percent. However, this idea, which emerged from the standard hardware world, soon crept into other areas. Software developers, for instance, wanted to know how many errors they could have per line of programming. There was a lot of time and money spent on negotiating this sort of stuff.

My response was that they had bid the contract to supply us with this service or product and I expected it to appear exactly as promised. There would be no AQL. The suppliers moaned that they were going to have to check things over and over; they stated that programming was an art, not a science; and they would seek relief from senior management. We held seminars to encourage them to learn how to conduct their processes based on prevention. If they did things right, then the end result would be correct. I would tell them about the hotel where the executive housekeeper created a perfect room and let the maids take photographs of it. This eliminated the follow-up expense of room checking. They would hear from suppliers of electronic components who had not found a nonconformance in millions of parts. All of this because the management was interested in getting the processes of their company understood and implemented.

We don't hear much of the economics of quality anymore. I think that has happened because customers have made it clear that they do not want to fool with having to do things over once they have purchased.

4. *The Belief That the Problems of Quality Are Originated by the Workers.* The general manager of an American automobile assembly plant once told me that he would not have a result of eight

defects per car if he had Japanese workers. I told him that he would have exactly the same result because the workers and staff were all performing to his personal standard. He did not accept this well. Management in this industry was convinced that the workers caused all their difficulties; in the past dozen years they have reversed that opinion and learned to work with their people. I campaigned to get rid of executive dining rooms, coats and ties, levels of supervision, and other customs that interfered with people getting to know each other. Today most of that has happened in manufacturing; service operations are just getting there.

However, service and administrative organizations are still trying to come to grips with this. Banks typically have a 40 percent or more personnel turnover rate at the lower levels. Hotels, insurance companies, restaurants, retail stores, and similar businesses suffer the same fate. Most of this comes because these employees have not been oriented, trained, and shown a career path. They are just shoved into the jobs with the thought that they won't be around for long anyway. It is hard to believe that executives still think this way. They need to concentrate on creating "veteran" employees.

5. *The Belief That Quality Originates in the Quality Department.* The situation as covered in *Quality Is Free* has corrected itself a little bit. Management and quality professionals both have come to the conclusion that quality is the result of actions, not something the beat cop can cause alone. However, now the feeling is that quality is the responsibility of a committee or team made up of people from around the company. They use ISO 9000, Mil-Q-9858, the Baldrige criteria, or the tapes of a guru as their guide. If finance were handled this way, the executives would be locked up for incompetence. During a recent trip to Europe and India I found a complete frenzy of companies trying to scramble into the ISO tent. It would be funny if it were not true that they were headed for a large disappointment. Improvement will not come, nor will a protective wall form. The next time around it will be even tougher to convince people that the management is serious about quality.

Companies which have been successful in creating a culture of quality have done it under the direction of senior management who have been assisted by someone who knew what they were doing. Putting a group of well-meaning amateurs in charge is like turning the operating room of the hospital over to the volunteers.

In spite of all the frenzied activity concerning quality since *Quality Is Free* was published, there is still no general agreement as to what quality is or how to get it. Because the subject appears to be subjective and easy to understand, a great many people have leaped on it. That experience will be covered in more depth when we discuss the new Quality Management Process Maturity Grid.

THE QUALITY MANAGEMENT PROCESS MATURITY GRID

Quality Is Free introduced the Quality Management Maturity Grid (see Figure 4-1). It offered a way of evaluating the status of a company's quality approach in that moment of time. It has proven to be a useful communication and analysis tool. However now we are in a new age, so we need to use a new evaluation tool. The Quality Management Process Maturity Grid (Figure 4-2) concerns the managerial actions that organizations are taking in trying to improve and the problems they face in making that happen. The creation of the original Grid came from the task of introducing managers and executives to the actions that were involved in running a successful quality management effort. Most of them were not familiar with the content of such an activity so it was necessary to lay it out in a way that they could quickly understand. The trip from Uncertainty to Certainty is designed to make the next step obvious. But its real value was to show that there is more to handling quality than was going on at that moment in the company. This helped to make executives receptive to the idea of taking a new look at quality as a management problem rather than a technical activity.

During the past 15 years most companies have realized that they have to do something organized about quality. The Management Grid has been reproduced in hundreds of company

MEASUREMENT CATEGORIES	STAGE I: UNCERTAINTY	STAGE II: AWAKENING
Management understanding and attitude	No comprehension of quality as a management tool. Tend to blame quality department for "quality problems."	Recognizing that quality management may be of value but not willing to provide money or time to make it all happen.
Quality organization status	Quality is hidden in manufacturing or engineering departments. Inspection probably not part of organization. Emphasis on appraisal and sorting.	A stronger quality leader is appointed but main emphasis is still on appraisal and moving the product. Still part of manufacturing or other.
Problem handling	Problems are fought as they occur; no resolution; inadequate definition; lots of yelling and accusations.	Teams are set up to attack major problems. Long-range solutions are not solicited.
Cost of quality as % of sales	Reported: unknown Actual: 20%	Reported: 3% Actual: 18%
Quality improvement actions	No organized activities. No understanding of such activities.	Trying obvious "motivational" short-range efforts.
Summation of company quality posture	"We don't know why we have problems with quality."	"Is it absolutely necessary to always have problems with quality?"

Figure 4-1. *Quality Management Maturity Grid. (Reprinted from* Quality Is Free, *copyright 1979 by Philip B. Crosby. Reprinted with permission.)*

papers, as well as magazines and books. The result has been to push management into installing quality improvement and management processes. Some companies have been successful; most

STAGE III: ENLIGHTENMENT	STAGE IV: WISDOM	STAGE V: CERTAINTY
While going through quality improvement program learn more about quality management; becoming supportive and helpful.	Participating. Understand absolutes of quality management. Recognize their personal role in continuing emphasis.	Consider quality management an essential part of company system.
Quality Department reports to top management, all appraisal is incorporated and manager has role in management of company.	Quality manager is an officer of company; effective status reporting and preventive action. Involved with consumer affairs and special assignments.	Quality manager on board of directors. Prevention is main concern. Quality is a thought leader.
Corrective action communication established. Problems are faced openly and resolved in an orderly way.	Problems are identified early in their development. All functions are open to suggestion and improvement.	Except in the most unusual cases, problems are prevented.
Reported: 8% Actual: 12%	Reported: 6.5% Actual: 8%	Reported: 2.5% Actual: 2.5%
Implementation of the 14-step program with thorough understanding and establishment of each step.	Continuing the 14-step program and starting Make Certain.	Quality improvement is a normal and continued activity.
"Through management commitment and quality improvement we are identifying and resolving our problems."	"Defect prevention is a routine part of our operation."	"We know why we do not have problems with quality."

Figure 4-1 (*Continued*).

have not achieved the results that have been hoped. The managers' hearts were in the right place but they were pushing buttons that were not connected to anything. For this reason I have

	Uncertainty	Regression
Management Concepts	"Let's get certified."	"Let's apply for the Award."
Definition	Goodness	Delight the customer
System	Award criteria	Buy some guru tapes and show them
Performance Standard	What traffic will bear	Acceptable Quality Levels
Measurement	Opinion	Benchmarking

Figure 4-2. *The Quality Management Process Maturity Grid. (Copyright 1994 by Philip B. Crosby.)*

constructed the Quality Management Process Maturity Grid. This again moves from Uncertainty to Certainty with three other levels in between. Its purpose is to quickly show management where they have stepped off the lily pads. The process of quality is not complex, but anything can be made incomprehensible if one gets involved in the wrong stuff. The mantras and chantings offered by gurus and other teachers should be held up to the bright light of reality. Go with what works, not with what is supposed to produce results. I see company after company diligently slogging along some laid-out path to improvement only to be disappointed. They are trying to learn the twists and turns of the road when what they need to understand is how to drive.

As management begins to be concerned about lack of change, I am often asked to "drop by" and evaluate a corporation's status concerning their quality improvement process and its results, if any. The managers who hire me figure that since I am not in the business of selling quality management education any more, I can be trusted to give them an objective view of the company's status and progress. I really enjoy these involvements, and I can provide pragmatic feedback, although it is often not what they are prepared to hear. Usually the process takes a few

Awakening	Enlightenment	Certainty
"We need to get better."	"Get serious about quality."	"No reason for not doing things right."
Continuous improvement	Satisfy customer	Conform the requirements
ISO 9000; Mil-Q-9858	What do we really need to know?	Prevention
Continuous improvement	Six Sigma	Zero Defects
Customer complaints	Complete Transaction Rating	The Price of Nonconformance

Figure 4-2 (*Continued*).

days, depending on how much the company is spread about. Basically, it is necessary to find out what management thinks is happening, what those in charge of implementation think is happening, and what is actually happening. Management usually takes my determinations very seriously and goes right to work on changing, provided something needs changing. If we were to select a dozen companies at random, I would predict that 10 of them would be spinning their wheels and 2 would have achieved a modest amount of progress, mostly anecdotal. They are all working hard and are sincere about the effort, but as kids say: "They just don't 'get it'." The winning football plays are those that overcome the defense presented to them, not those that have a brilliant design. When the coach understands the defense, the team can be instructed to implement a certain offense. This strategy requires that players understand the objective of the play—and recognize their specific role in executing it with zero defects, including zero penalties.

Typically a call requesting my visit will come from a CEO's office and I will make an appointment to meet with him or her for a chat, usually over lunch. Most of the time the meeting happens in or near the CEO's office since I live off the beaten path, but often the CEO comes to Florida or North Carolina. Once in a while this meeting turns into a golf game. If you can't get to know someone during a round of golf, then there probably is no chance it will ever happen. The CEO always begins conversation with a summary of how many wonderful things have been happening since the company got started in this process. If the CEO refers to "TQM" and talks about activities, then I begin to be suspicious of that person's comprehension. All of these questions enable us to determine at which stage of progress, from Uncertainty to Certainty, the company is at. The progress can be marked on the corresponding box on the Grid. Usually "x" marks a spot that is not what the CEO had been led to believe.

But the purpose of getting together is for me to understand and help clarify the CEO's concern. The meeting would not have come to pass if everyone were contented. Usually I will be given these concerns early in the conversation, sort of what one does with a marriage counselor. As with marriage counseling, each person knows more about how others are behaving than about what he or she personally is doing. The thought that that person might be the problem usually does not compute early on. So eventually the story comes out that a lot of wonderful things are happening. Teams have been organized and are making helpful actions, awards have been won, tales of accomplishment by individuals are heard all the time. Other companies visit in order to see what is going on and are complimentary about what they see and hear.

However, the boss wonders. There is still the usual amount of problems coming to this office; there seems to be no effect on the bottom line; there is still little confidence that all of this change is going to continue. Usually these concerns are legitimate. The only person in a company who looks at the whole picture is the CEO. Everyone else has a private agenda which they may not reveal to anyone. If the boss thinks things are not as advertised, then you can bet that this is the case.

The purpose of The Quality Management Process Maturity Grid is to help all employees "get it" by understanding what goes on inside an organization when they decide to do something about improving quality. Most efforts fail, and the matrix will help us realize why failure is more frequent than success. Those responsible for driving the process can use the Grid as a measuring stick.

The Grid only works when taken literally, without rationalization or transposition. Several of my friends who died in what should have been the prime of their life had the same story of an uncle who was heavy, drank like a fish, ate fried food plus whatever else he wanted, smoked three packs of cigarettes a day, and lived to be an active 95. That uncle was a statistical error, as their fate showed. Damon Runyon wrote a story about some tortoises who read the fable about the race with a rabbit. They took bets from everyone and got set to make a killing. The rabbit was across the finish line before the tortoise left the starting line. The moral? "The race is not always to the swift, but that is the way to bet."

At the time *Quality Is Free* was published there were very few formal quality improvement efforts going on in business. By "formal" I mean something blessed by management, having an agreed agenda, sporting agreed goals, and following some sort of process. It is the "process" that matters. Quality was always viewed as a "program" and we have learned that the casual use of that word sends people off in the wrong direction.

There was a lot of confusion about how to do something about quality back in those days. The organizations with established quality-control departments were working on getting better, but the responsibility was left entirely to that function. It was like holding the facilities department responsible for getting people to park their cars between the white lines exactly, and directing them to get twice as many cars into the same amount of space. It is easy to just paint over the parking lot, halve the size of the slots, and thus meet the space objective. The employees could then be assigned blame for not being able to carry out the program. This is a silly example but it is a correct analogy when it comes to quality improvement processes. There was little realization that the entire company caused the quality situation. But 1979 was the time that awareness was beginning to set in.

I had been complaining about the American attitude toward quality for 20 years and receiving little positive response. It wasn't until quality became an obvious concern of the customers that attention began to be paid. Japanese and German products worked better and offered more reliability than the consumers had ever imagined. American managers and their quality professionals had been hung up on the idea that it cost more to make things right. That is one of the reasons I called the book *Quality Is Free*. As I noted, no one liked the title. The phrase originated with Harold Geneen, the impresario of ITT's spectacular growth.

We installed a successful quality management process (not a system) in ITT, worldwide, for a dozen years, as the opening chapter of the book related. ITT suppliers had been encouraged to do it also and reacted enthusiastically. Usually they were the first in their industry to tackle quality management systematically. Today everyone is doing something. That is progress, of course, but doing something in order to do something, and doing something that causes positive and permanent change are vastly different.

Today we hear companies talking about their quality management processes and relating the success or lack of it that they are having. Awards are being given, articles are being written, tours are conducted, and it would seem that everything is going well. However, many people confuse activity with results. Overall there is disappointment and turmoil; a lot of people are disappointed with the results. There is nothing complicated about the concepts of quality management. Causing them to become part of a company's culture is not difficult or expensive. But it is hard to know where one is at any given moment. The Grid will tell when no one else will. It should be possible to determine where an organization sits at any given moment and go on from there. It is not necessary to be exact, just to know what bad work is going on and exchange it for good work.

I might recommend The Wellness Grid in *The Eternally Successful Organization* (McGraw-Hill) as a way of looking at the existing status concerning operating practices. It was created to be a second-generation Maturity Grid.

UNCERTAINTY

Uncertainty has always seemed to me to be like something freshmen undergo when planning for college. Having no experience along that line but having heard a lot from many sources, they put together a scheme that does not deal with reality. They know nothing of reality. All this begins to show up in their second year when they switch major study lines, arrange to room in different places, alter their eating habits, and even change colleges. Uncertainty can be examined in more detail by looking at people's actions in organizations in comparison to the Absolutes of Quality Management and the attitude of management.

MANAGEMENT ATTITUDE: "LET'S DO IT TOO"

This is an idea-of-the-month mentality where executives look at the surface idea and launch an effort to show that they are up-to-date. This is like a student signing up for pre-law in order to show cousins how smart she or he is. After living with the courses for a few weeks and deciding that "communications" would be a better way to go, the student clumsily backs out. No one may ever quite trust that person's plans again. With the subject of quality, management gets wrapped up in the media material in which writers who have no clue about reality explain it to those eager to be modern. Many get locked into one "guru" and assume that this is where all wisdom is centered. The only real things a guru can teach is something they have experienced. Most of them have never run anything. "The Japanese quality model," for instance, exists mainly in the imagination of the media.

DEFINITION

With Uncertainty, quality is thought of as "goodness," which is a matter of opinion. During this stage, the common thought is that doing things right means doing them slowly, involves a lot of checking, and requires a reduction in innovation, productivity, and creativity. With Uncertainty, the subject of quality has never been given a good thought.

When I first dealt with an American automobile company, in 1967, this stage was as far as we got in terms of improving automobile quality. The management was totally convinced that cars had to be made as fast as possible and then fixed later; they wanted the workers to have as small a job as possible; and they felt that quality was expensive.

The hotel I stayed at during this visit had gotten the same beliefs. Their way of quality was to have a checker who came around unannounced and inspected what was happening. Things that were not "good" were reported, and the worker involved was disciplined. They advertised themselves as a "quality" hotel. They should have called it a "corrective action" hotel.

System: "Award Criteria"

The CEO picks up the criteria for the Baldrige or NASA or some similar award, hands it to someone, and instructs that person to "go do this." Several dozen pages long and full of proper-sounding verbiage, the criteria gives the impression of containing the wisdom necessary to install quality management. Actually the material represents what was being done in the quality assurance days of the 1960s and 1970s. Those were the golden days when acceptable quality levels were standard and nothing worked well at all. These systems didn't work then and there is no reason to suspect that they will do so now. The only result will be a few books full of procedures which will sit on the credenza of selected managers.

Performance Standard: "What Will the Traffic Bear?"

This group reacts only to customer complaints and problems that threaten them. If no lawsuits or horse whippings are forthcoming, they assume that everything is in good shape. The employees receive their standards from each other; they know that management has none. This is the mentality that originated the cliché "good enough for government work." If asked to state their performance standard, the employees gaze blankly.

MEASUREMENT: "OPINION"

Never have any measurements with numbers in them, and do not seek such things. Daily activities are punctuated by arguments between those who have to decide on the work in process. Each step of the way there is disagreement on the status of the material or service. Those with the loudest voice and sturdiest stance usually have their way. The reason things go on like this is that management is not interested in numbers that do not have dollar signs in front of them.

REGRESSION

After the first half-hearted swipe at quality improvement goes nowhere, management stirs itself to look for some external happenings that will make a difference. They look for activities and events that will get the attention of those who are complaining. I see these folks as dieters who are searching for some way of eating whatever they want, doing no exercise, and still losing weight. The only way to make that happen is to juggle the books.

MANAGEMENT ATTITUDE: "LET'S APPLY FOR THE AWARD"

The Wizard of Oz solved the Scarecrow's problem of having no brain by giving him a diploma. The Tin Man got a heart to wear around his neck; and the Cowardly Lion received a medal for bravery. They all went away contented. It is the same with this management attitude. If managers can figure a way to win an award, then all the problems of quality will slip away. They hire a consultant to fill out the forms, they get procedures written to show compliance, and they make up a brochure to apply for consideration. It all seems like positive effort. It is like finding a doctor who fixes the X rays rather than doing surgery on you.

DEFINITION: "DELIGHT THE CUSTOMER"

I find it hard to come up with any definition that has less value than this one. Imagine yourself standing on a box in front of 500 employees and instructing them to go out and "delight the customer." There would probably be a few sexual harassment suits as a result, if people took that literally. The way to delight the customer is to determine their need and then work hard to create the requirements that are necessary to meet that need. It is not a bad instruction to give to the creative departments, but they are abstract enough to appreciate what it means.

SYSTEM: "BUY SOME TAPES AND SHOW THEM"

There are catalogs of quality tapes, and some of them are actually worthwhile. However, setting the employees in a conference room and showing them a few hours of tapes changes nothing. Imagine trying to explain boys to girls or girls to boys by such a method. It takes a much more personal involvement. One company bought thousands of paperback copies of *Quality Is Free* and distributed them to their employees. They were disappointed that a tremendous surge of quality did not pour forth. I told them they should have bought hardcovers instead of paperbacks.

PERFORMANCE STANDARD: "ACCEPTABLE QUALITY LEVELS"

When I started in the quality world in 1952 my first lesson was about AQLs. There was scientific proof that it cost a lot less to plan on things not being correct. "It would take a fortune to do everything right every time" was contained in every explanation. So all processes were planned around these statistics. It was a great performance standard because we weren't expected to do it right and when we didn't there was no problem. Suppliers had AQLs included in the purchase order so they could have a defect rate of 1 percent, 2 percent, and such. Sloppy executives love AQLs.

Measurement: "Benchmarking"

It is always possible to find someone fatter, uglier, and dumber than you are. For that reason most Benchmarking is dishonest. Here is a good idea gone wrong. Certainly we must keep up with what the competition is doing, but that doesn't mean it is permissible to sink to their level.

AWAKENING

After much discouragement and expense, management begins to notice that nothing is changing, except the people are probably becoming more cynical. After hearing how wonderful everything is going to be, and watching the pretend effort going on, the employees tend to lose faith. Often this is the time that a new management team comes into the organization. However, there is nothing wrong with the old team awakening; it can be done. My experience with large companies, both manufacturing and service, is that they build up so many traditions and customs that the current management just can't change. They have entrenched beliefs, and to their credit many can throw them off, but it is difficult. It is the equivalent of having a revelation in church and walking down the aisle to state that one has not realized they have been working on the wrong things all this year.

Management Attitude: "We Need to Get Better"

So what do we do about it? At this stage enthusiasm usually overcomes the need for learning. A team leader is appointed and everyone starts reading what the media says about quality management. This is like becoming one's own medical expert and taking a dozen drugs each day with the hope that all the varied conditions will be cured. When an organization has problems with quality, there are a lot of causes but the treatment is much the same. Senior management should find a company they respect that has done well with quality improvement as reflected in market share and profitability.

Then the CEO of the first company should go see the CEO of the second and find out what they did. Don't send a quality committee leader to go talk to the quality committee leader. They usually don't have a clue as to why things really work.

Definition: "Continuous Improvement"

People get angelic looks on their faces when they talk about continuous improvement. There are actually two different concepts involved. First is the popular one that says it is okay to drop six babies this week as long as we only plan to drop five next week. The second is that once we learn how to do things right, we are going to learn to get better all the time. Very few like the second. It is a lot of work. The first concept permits a group to drag along doing things that don't cause much pain or discomfort. They can talk about their noble objective. But it is like golfers who take lessons, buy tapes, get new clubs, and play a lot but never improve. The problem is in their heads, not in the equipment.

System: "ISO 9000; Mil-Q-9858; Baldrige Criteria"

Wouldn't it be wonderful if there were a system of personal success that came in a 40-page package, with notebooks and procedures? All we would have to do is follow the program and we would wind up with a great job, a big house, a charming family, and tons of money to invest. We all know personal life is not like that. Neither is quality life. There are no good fairies who will change a mushy outfit into a tough company. Why otherwise sensible people would believe that some group of "experts" can lay out a process that will set their company free of problems is beyond me. Most executives who command their people to comply with these specifications have never read them.

Today, often customers will ask that their suppliers become certified to ISO 9000 (it was developed by a European group whose previous experience was concentrated on standard hardware). With a properly run quality management process, there will be no difficulty meeting ISO 9000 requirements. It is really a very old-fashioned Quality Assurance kind of thing. But it is not oriented toward the needs of today and the next century. It is only to provide a living for consultants who certify companies and for quality professionals who do not want to think for themselves.

Performance Standard: "Continuous Improvement"

As noted earlier, in Awakening this standard means that you never really have to get better, you only have to be trying. This is an example of a cliché gone wild.

Measurement: "Customer Complaints"

What finally prods Awakening are stern customers who say that they are tired of fixing what they receive: having to debug software, shine nicks out of metal, put salt on potatoes, or a lot of other things. The requirements that were promised the customer have not been met. A scurry of activity produces better results, and the customer complaints reduce dramatically. At this time management often breathes a sigh of relief and goes on to other matters. This means the complaints will return, of course.

ENLIGHTENMENT

When we finish school, marry, and set up our own home we realize for certain that we are responsible for our own lives. It is up to us to provide the cash flow, to mow the yards, to discipline children, to make all the choices that mature life requires. This is Enlightenment. In business executives this means that they realize quality can only come from their personal efforts. They have known this about profitability; they have known this about marketing perhaps. But they have been led to believe that they can "hire quality done." They should know, from trying to get someone to cut their yard properly, that no one does it as well as the principal person. That is because the principal person is the only one who looks at the whole picture. Everyone else has a much more narrow view.

Management Attitude: "Get Serious about Quality"

When I set up the Quality College in PCA in 1979, its purpose was to instruct executives and managers about their role in the quality process. I purposely wanted to deal only with those who were ready to enter the Enlightenment stage. So there were no sales calls, every client came on their own. As a result they knew that they were going to have to do the work once we explained it to them. There would be no hearts issued to wear around the neck. We did give them a plaque for their desk so they could explain it to every visitor, but that was a marketing idea of mine.

"Getting serious" means abandoning hope that there might be some canned wisdom or programmed steps that would revise the working culture of an organization. It means getting real. When I think of this, I remember when I first became a husband and then a father. My family would present me with some problem or opportunity that faced us. I soon realized that neither my parents, neighbors, government, company, or anyone else was going to deal with it. I had to do something about it or it would not be resolved. When that enlightenment became obvious I took action and did it with purpose and efficiency. This might have involved getting a leak in the roof fixed; dealing with a neighbor's child who was bullying my son; getting a new set of tires; smoothing a set of hurt feelings; making more money; or a million other situations. The important thing was the decision that nothing was going to happen unless I made it happen. Each of us goes through our own coming of age in some fashion. The calls we received at PCA from senior executives all started out the same way. "We are just not getting anywhere on our own," they would say. "Can you help us?" If there is one action that typifies enlightenment, it is the recognition that it is acceptable to ask for help.

Definition: "Satisfy the Customer"

This goal is not a complete one but it is on the right track. "Satisfy" is not a specific enough word to give people direction, but it does help them get the idea that we are supposed to produce what the customer wants. In philosophical terms it is a flip response that implies everyone knows what it takes to make the customer happy. Those who run quality programs like it. Those who are measured on results do not. I prefer "successful."

System: "What Do We Really Need to Know?"

Tossing aside packaged activities, a company can now approach the situation with questions rather than answers. The list begins like this:

"What is our problem?"

"Where are the causes?"

"What caused the causes?"

"How can we learn to prevent these causes?"

Performance Standard: "Six Sigma"

At this point we are not happy about errors but we understand (incorrectly, of course) that such things are going to happen now and then. Therefore, the company will have built-in rework activates such as an assistant manager sitting at a desk in the hotel lobby to handle guest complaints. It is necessary to take the great leap over the chasm of accepting error even a little bit, and this is a beginning. Those doing the leaping think they have done a great event and do not appreciate the downplaying that I usually give them. One group pointed out their transformation to me and, before thinking I said that they were the "best of a bad lot." They printed that out on a huge banner and hung it in their conference room. After they had moved into Certainty they kept it there as a reminder in order to stay there.

Measurement: "Complete Transaction Rating"

The speedometer on an automobile is a statistical analysis tool. Its purpose is to let the driver know what is going on; it has nothing to do with causing driving skill. This is a distinction that seems obvious but is not often made by those who are enthusiastic about generating numbers. They soon become comfortable with manipulating numbers to their own advantage. When the minimum speed is 40 miles per hour and the maximum is 65, we have established the limits of our SPC chart. However, when it is necessary to get to the airport on time, the manager may decide to exceed that limit. There is always a story to go along with it. This same thing happens with business statistics when they interfere with an immediate goal. The next chapter talks about the Complete Transaction Rating (CTR) as a measurement. Next to money it is the best one, and provides a good compliment to the price of nonconformance.

CERTAINTY

To be certain means that we know what is going to happen; that is not a common occurrence in life. We could be willing to settle for knowing what is not going to happen. This stage assures that the members of an organization are properly committed, educated, trained, and led. Maintaining and improving that situation then becomes the concern of management. It is hard to win the championship of a league two years in a row because the management does not take proper action to keep the team at its best. Success seems to cause people to take their eyes off the things that are really important. For that reason it is necessary to rethink the actions that brought the company to Certainty and do them over again, perhaps in a different way. One can never acknowledge being finished. However, very few ever reach Certainty; they are not willing to work hard enough. In Certainty the constant Complete Transaction Rating is 1.0, people take their work very seriously, and are proud to be part of the organization.

MANAGEMENT ATTITUDE: "NO REASON FOR NOT DOING IT RIGHT"

No excuse is acceptable. Everyone has the opportunity to help create the requirements of their job; communication channels are forced open at all times; and there is no harassment.

Definition: "Conformance to Requirements"

The old days of quality being a vaguely defined entity are gone. Each person in the organization, including those suppliers who deal with it, know that management expects things to be clear and clear things met. There is a common language.

System: "Prevention"

This is more of an orientation than a system. The idea is to prepare properly before doing something. Instead of setting out on a world tour with a handbag and $10 in cash, we are going to plan the whole adventure carefully. We do not want to fly into a dead-end canyon, or walk casually out toward the horizon in Death Valley. Building a culture of prevention raises the intellectual level of the company. It generates ideas and actions; it eliminates feelings of repression because it focuses everything on what is coming. It is hard to shoot yourself in the foot when the gun is aimed forward.

Performance Standard: "Zero Defects"

To attain a CTR of 1.0 it is necessary to do things right the first time. To cause that, management has to make it clear that they value this sort of behavior. More sports teams are victorious because of their opponent's errors than because of their own efforts. The team that has the most turnovers usually loses. The individual performer in business, sport, or personal life that develops the reputation for complete transactions gets ahead. Those who are always trying to snatch victory from the jaws of defeat are no longer valued.

Measurement: "The Price of Nonconformance" (PONC)

The amount we spend for the purpose of doing things wrong is a great deal of money. Many companies have difficulty coming to grips with this measurement. Quality professionals often feel that it makes them look bad. Accountants can't find it in their books, and top management does not insist often enough. But PONC will become common as executives realize the reality of the money pouring down the drain.

This is the reason I like the idea of CTR. When you become transaction-oriented the obvious costs of doing things over become apparent. A rough way of figuring a CTR is to determine how much a task costs to have done in the first place. For instance, to fit someone incorrectly with a pair of shoes means that the customer will come back to get the right size. If it takes 20 minutes to fit and sell shoes and we have to do it over, plus handling the return, then we need to charge ourselves 40 minutes of money. The extra 20 minutes comes from the lost sales that could have been taking place at that time. It is easy to determine each person's CTR when we become familiar with the concept.

USING THE GRID: A CASE EXAMPLE

The Smedley Corporation is a 220-person company with revenues of around $24 million and an after-tax profit of $1.2 million. Their product is considered a commodity and their customers relate to price and service. Smedley is a privately owned company that would like to go public. Their profitability has never been robust enough to interest a brokerage firm in arranging the transaction.

"If we could raise our margins to where we were 20 percent more profitable, we would be able to go public in a flash," said CEO Nancy Robinson. "But with our competition as it is now, and with Latin America entering the field, it doesn't look too hopeful."

"What would you do with the money you would get from an IPO?" I asked. "Would it make you more competitive?"

She nodded.

"There is a whole level of new manufacturing technology available to us if we could pay for it. Also we could buy into a few of those Latin American companies who will be competing with us. As the product line grows and broadens, we can enter the Far East where there is very little opposition at this time."

"Do you have projections and all that stuff?" I wondered. "I took a company public once, about the same revenues as yours, and I never got over the amount of paper involved."

"We can handle that. What we need is to get our costs down. That is why I asked you to drop by," she said.

"You plan to get this cost elimination out of quality improvement?"

"Not directly, but I know that we must be wasting a lot of money. Our cost of manufacturing has risen higher than our efficiency improvements can handle. We have a quality program in place and it is getting great results. But I wonder why we don't see more of it where the money grows."

"What would you like me to do?" I asked.

"Perhaps you could look at what we are doing and give us some ideas about how to get more results. Our customer complaints are a little less than they were this time last year, and our field service people report that customers trust that we will take care of them."

I nodded and pushed some of the notebooks she had given me around a bit.

"Did you have a chance to look at the Quality Management Process Grids I sent you?"

"Yes, I have them right here. I sent them to my key executives and managers with a request that they be marked and returned without discussion."

She handed the stack of seven pages to me.

"Is yours here?" I asked.

"I didn't do one, I thought it was the executive staff we were concerned with so I didn't even look at them. But give me a blank and I'll do it right now," she replied, a little uncomfortably I thought.

"There is no need," I commented. "Let's just discuss the quality process you have going. How do you think about it over all? How did you describe it to the board, for instance?"

"I don't remember mentioning it to the board, it didn't seem like something that should involve them. But I guess I should do that. My comment would be something like 'we need to do quality management formally in order to compete with others'," she smiled and nodded her head.

"What system would you use to do that?"

She pondered a moment.

"Well there sure is a lot written about TQM and quality in general. What we did was appoint Charles Opten to be in charge of quality and he has put together our own system based on what others have done. He passed around a set of papers from the companies that won the Baldrige award, just last week."

"Did you read it?" I wondered.

"Not yet but it is in my weekend stuff."

I turned to the grids she had handed me, with one exception they all had marks concentrated in the Uncertainty and Regression areas. The exception was a straight line down the Awakening slot. It was signed by the quality manager.

"Your executive staff seems to feel that the company is in bad shape from a quality standpoint," I noted. "Perhaps we should start out with putting together a list of our problems before we start working on systems to resolve them."

She smiled.

"That makes sense. Perhaps we have mounted a horse and galloped off before we know the destination. What would you guess will wind up on our top list?"

"If you want me to peer into the future I would say that there are indeed five areas that will need prompt attention. They are also the ones where waste is prevalent."

"I was teasing you when I asked that," she said, "but if you really are going to do that, why don't we wait until after lunch and we can gather all the management together. That way they can let you know if you guess right."

"Sounds good to me," I responded. "However very little guessing is involved. You all provide the answers."

After lunch the staff gathered together as promised and Nancy introduced me to each of them, providing a brief description of their responsibilities at the same time. She recounted our conversation and said that I had promised to list the five major items we should be working on.

"Don't we need to get our quality program laid out and operating before we take on some specific tasks?" asked the finance director. "I have read that it takes several years to prepare the way for real improvement."

I smiled at him and shook my head.

"The nice thing about management is that they are usually the cause of the problems and so they can make the necessary changes in a hurry. You can turn this company around in a few months. I figure that your price of nonconformance is around six million dollars. If you could reduce that by 50 percent you would not have to spend the time and money necessary to go public. You could use your own money to expand and even find a proper merger partner, if you wanted.

"Anyway here is my list:

"First, since you have no definition or policy for quality there is constant quarreling about the products and whether they are good enough to deliver to the customers;

"Second, your field service people spend all their time repairing and explaining instead of creating new customers;

"Third, most of the supplied items you receive do not conform to the purchase orders and you have to repair or discard them. You have a great many suppliers since all orders are given to the lowest-priced response;

"Fourth, employee turnover, particularly at the lowest levels, is very high, probably 25 percent a year. Also many people are absent or tardy on a regular basis;

"Fifth, you have checked out other companies who compete with you and find that they have the same situation, so you feel that you are in good shape."

There was a long pause. Each of them had been writing these points on their note pads but now they sat and stared at me.

"You have been spying on us," smiled the manufacturing director.

"What do we do about all this?" asked the CEO. "Will putting in TQM solve these problems?"

"That would probably just make it worse," I replied. "All you need to do is to take action to change the way you run your company. Business is complete transactions and successful relationships. Those can only be generated by a management team who seriously want them to become routine. Look at the Grid under "Certainty" for a guide.

"If you as the management take the stand that we are going to learn how to do things right, issue a policy on it, and then provide some training to support it, change will begin.

"If you define quality as conformance to requirements, then you eliminate the 'is this good enough?' argument. This will result in eliminating the need for field service repair and apologies. That area can be used for growing customers.

"Relationships with suppliers develop over a long time of doing business together. You don't need all those suppliers. Select some that will take good care of you and develop a relationship. The effect of this will be felt right through the whole process. This is part of the prevention process.

"Employees don't stay around when there is confusion and little emphasis on training. They don't like to come to work when nothing ever gets better. They don't take the company seriously if they feel that management does not do it either. When they see you taking action to prevent problems they will change their attitude.

"Benchmarking to learn what is happening is a good idea, but doing it to see if anyone is better than you is not. You can be quickly satisfied. Just assume you are far behind, which you are, and learn to improve. Zero Defects is what you need especially if the competition is using Acceptable Quality Levels."

"Start measuring the processes by setting up a Complete Transaction Rating; and measure the company with the Price of Nonconformance. These are real tools that will help you manage.

Nancy squinted at me.

"You got all that out of the way we checked off the Grid? You haven't been anywhere but my office."

"Well," I agreed, "that is true, but you did show me some figures and I did get to meet three employees out in the reception room.

"However, a great deal is obvious from the Grid. You all rated yourself in the Regression stage which is where there are no clear definitions, where quality is not taken seriously, where there is very little training, and companies compare themselves with each other instead of looking to the customer. Some is guessing, but that comes from experience.

"You will do all right if you point yourself at the right tasks."

I will spare the reader the rest of the conversation, but this group did take all that to heart. They got serious about quality by establishing a policy, implementing an education and training program, and serving as proper examples of doing things right. The relationships with employees, suppliers, and customers improved quickly once they started paying attention to them. The business grew 20 percent in revenues without the need to add employees, profit rose to 8.2 percent after tax, and employee turnover dropped to a trickle. All of this took place within six months.

These folks had been looking for some magic system, installed by a specially appointed person, to save them. But they had all they needed right at hand; it was just a matter of putting things in the proper perspective. That is the value of the Grid.

THE COMPLETE TRANSACTION RATING

The bigger bugs have smaller bugs,
To sit on their backs and bite 'em
The smaller bugs have littler bugs,
And so on ad infinitum

ANONYMOUS

As I noted in a previous chapter, business consists of transactions and relationships. This is not just a casual phrase to toss off; these two subjects are really all that happens in this complex world of commerce. We do transactions for people and they relate with us based on how well we complete our agreed tasks. People will not do business with some organization or person they consider unreliable, unless they have no option. The world today provides options on just about everything. It is not possible to obtain a different biological mother or father, but outside of that nothing else seems unobtainable. The divorce rate in most nations is around 50 percent because people decide they have made a mistake with the original product. If we listen to the stories couples relate concerning their problems, they are all wrapped up in transactions and relationships. The majority of small businesses fail and the reasons are the same: incomplete or failed transactions and relationships that are not successful. It's not money; it's the doing.

It is difficult to explain concepts like transactions and relationships today in one business example because people are usually oriented only to their product or service field. An example from some other aspect of work does not turn them on. In fact, they reject it. I have to be very careful when making a speech or discussing something with a group. If I say anything that sounds like a product, the service people immediately reject the rest of the sentence. Manufacturing devotees do not accept anything that doesn't include some sort of widget. They seem to feel that everyone is different. Actually every business operation is really much the same. Those in the service business have products: insurance policies, omelets, cars to rent, hotel rooms, and so on. Manufacturing people have service: salespeople, contracts, accounting, financing, advertising, and such.

If we think of a company that takes seven or so components, places them in processes requiring heat and pressure, packages them in a box, and then delivers the resulting assembly to the customer, most folks would say that is manufacturing. Of course it is, and the company that performs this process billions of times is McDonald's. That is the organization people usually list first when I ask for the name of a service company.

We could select one transaction that is common to every business. This is answering the incoming telephone call, or fax, or E-mail, and dealing with the caller's request. The successful company can do that every time, the first time. Not many are successful in this regard. The callers usually have to have great patience and diligence if they are to obtain answers that will satisfy their needs. Just getting through the voice mail answering system can be a challenge.

Recently I called a company to return a call I had received from its president. A voice mail thing answered to tell me that I had several options: if I knew his extension, I could dial it; if I didn't, I could spell the first three letters of his name and press #; there were several other instructions on the way, so I hung up. When I tried again I got through, to his voice mail box. Now who is all that stuff for? Are they saving money, like the salary of a telephone operator or two? People who have a choice will go where they can talk with someone. An experience like this cripples relationships. There is enough hassle in life without seeking it out.

And when real people are reached in an organization do they know how to help the caller? No one knows everything about a business, but they should be able to at least help the caller by putting her or him through to the one who does know. They should be able to do this while building a relationship in the process. Anyone who thinks their organization is in good shape should call in as a customer or supplier and see whether all of this is helping or hindering the company's objective.

Getting the phone answered promptly and properly requires more than gathering the staff together for a stimulating lecture. This activity is the top of the pyramid, as it provides a reflection of the company's attitude toward its callers, and it makes a clear statement on self-worth. Answering must be well thought out. I always told customers that I would guarantee that the phone would be answered before it rang three times, and that the person who picked it up would make certain their request was answered promptly and properly. Our associates took this seriously.

Transactions are the foundation of relationships. Recently I decided to trade my old car for a new one. I had decided on the type I wanted and that required me to deal with a dealership that was new to me. I did know the owner of the store from having served on a bank board with him. When I called the business and asked for him, the operator transferred me to his secretary who explained that he was on vacation, but asked if she could help me. I said that I was interested in buying a car and would come in next week, as I was out of town at this moment. She suggested that she put me in touch with the sales manager, and promptly did just that. He asked me what car I had in mind and I told him. The model year was just changing, he said. He had one new model and three of the current year. He described them at my request and I said that the new one was what I had in mind.

We arranged for my secretary to take my car over to him to be evaluated. The next day he called me, in Montana, and laid out the deal. I said it sounded right to me and that I would come get the car the following Friday. We arranged for her to bring him my current registration papers. When I arrived back home, I cleaned out my car, got the title out of the safe, stuck the checkbook in my pocket, and went over to the dealership. He showed me the car and insisted that we drive together to the gas station where he filled it up with fuel. He wanted to make certain that this was the car I had in mind and that I understood how it worked. I assured him that everything was correct. Then we went back to the finance office where I wrote a check for the difference in agreed worth between his car and my former one and signed enough papers to purchase a house. After that I spent 25 minutes at a desk and in my new car with another salesperson who took me through all the aspects of the vehicle from how the roof came off to the Florida Lemon Law. He had me initial alongside each item to make certain that I understood it all. He introduced me to the service manager who assured me of his eternal devotion. When I drove away they all came out to wish me well, and reminded me that they have a free "hand washing of the cars" every Saturday morning for all their customers. I've gone back for that, but the car has had no problems at all.

Now to make all of this happen there were hundreds of transactions—probably thousands, when you count what was necessary in designing and building an automobile that worked like it was supposed to work. The development of requirements for the actions of the staff, the instructional material that goes along with the car, and the training of everyone to make it all come out right represents a genuine commitment by management. Customers have come to expect this kind of service. Buying a car used to be rated, at least by myself, on the same level as a visit to the dentist. I liked the result but the relationship was painful. You can't see what either of them is doing to you.

We need to learn to view the actual act of handing the customer our completed transaction as the very point of the business pyramid. The little sharp tip of that pyramid is the act of depositing that person's check in our account. The great mass below that peak consists of thousands of actions that have to be done in order to receive payment and even gratitude. Trying to control that point by point is what caused the problems of quality in our business society. Causing everyone to know what they are doing and then helping them to go ahead and do it is a full-time job of prevention. Measuring, auditing, examining, and such have their own rewards but usually serve only to slow down the process while robbing the employees and suppliers of their interest in the matter.

The purpose of quality management is to build an organizational culture in which transactions are accomplished completely, efficiently, the first time. Transactions range from having the telephone answered before the third ring and the caller's request met, to the concepts of a new product or service. There are hundreds of thousands of them, each constructed on millions of specific requirements, in even the smallest company. Let's take a look at how it really works.

TRANSACTIONS: A CASE EXAMPLE

Helen Trant removed her purse from the desk drawer, tidied the area a bit, put on her suit coat, turned out the light and left her office, closing the door behind her. As she walked down the hall she checked to make certain the debit card was in her wallet. Her plan was to stop at the mall bookstore on the way home and pick up two books she had ordered. The store had called that morning to say they had arrived. She had only asked for them two days ago. The store she had used previously would have wanted three weeks to obtain the books. By then she would have had to find some other way to get the information.

Just before going out the door to the parking lot Helen glanced into the media room and noticed that Carl Turnburry was working at the conference room table. She stuck her head in the door to say goodnight but entered the room when Carl waved frantically to her. Hurriedly glancing at her watch, Helen walked over to the table. Carl was gazing helplessly at a pile of charts and printouts. He waved his hand at them.

"This is everything we know about our suppliers," he said. "I printed out some of the charts. Do you realize that we deal with over 2500 different companies? We buy computer software, we buy toilet paper, chemicals, insurance, travel tickets, the list goes on."

"I think I knew that," nodded Helen, "Is that what you dragged me over here—to let me know? We spend half our money on purchasing. So do most companies."

Carl shook his head.

"I know you know that. My problem is that George asked me to find a way to determine which ones are the good suppliers. They would like to cut our list down to the minimum."

"I thought you were working with a TQM team on all that. What happened to them?"

"Oh they are still at it but getting nowhere. They have some quality assurance procedures where we would go examine suppliers and do a statistical evaluation. We would never get done, not in my lifetime anyway. All that quality stuff has been no help; it just postpones decisions."

Helen looked at some of the charts.

"These show problems and the actions taken to correct them, and every supplier has some sort of rating number beside their name."

Carl grabbed several more papers and waved them at Helen.

"We have evaluations and calculations till hell won't have it," he growled. "Every one of our suppliers has a record of performance. The problem is that it doesn't make any difference. There is no problem that can't be explained away. The purchasing people have no real basis for turning anyone down. We have had three consulting firms look at this and they have nothing better to offer. Everything is after the fact, and imploring suppliers to do better."

Helen glanced at her watch again.

"So you called me over as a last resort? I only have about five minutes left. My book supplier is going to close before I get there. If I give you a solution in those remaining four minutes will that make you happy?"

"Thrilled," grumbled Carl. "What do you know that all these people don't know?"

"You know it too, all these data just cover it up. You are making it too complicated. Write this down on that flip chart.

"What are the components of the purchasing transaction? There is price, schedule, and quality. Each of these has components. What we want are suppliers who perform to each of these components exactly as agreed; their effort should be complete. So the way to measure a supplier is by setting up a complete transaction rating. Call it CTR if you want. If a supplier does it all completely, they get a rating of 1.0 and stay with us. Anything else, and they can work with some other organization."

Carl laid the papers back on the table.

"But the evaluation system the group developed is a dozen pages; you're talking about one number. Can we get by with that?"

Helen smiled at him and took a book from her bag.

"I changed bookstores because of it, and I deal with all my personal suppliers the same way. I don't have the time or expertise to examine what they give me before I use it; I don't have time to chase them around when delivery time is here. I expect them to know their business and be interested in my patronage. That seems to be a reasonable arrangement. In return I pay them promptly and recommend their services to my friends. This is coprosperity sort of thinking, where we are good for each other."

"So the idea," mused Carl, "is to get the variation out of the equation. Let's just deal with the cold, hard facts of complete transactions and the good relationships that develop from being complete."

"And the bad relationships that develop from the relationships not being complete," reminded Helen. "I buy three or four books a month. The store I had been dealing with just didn't take my requests seriously; there was always some problem. These new people just can't sleep until I get my books."

Carl was running the list of suppliers down the computer screen.

"If I ask for a call-up of the suppliers who have never had anything returned, or against whom there has never been a complaint from our people, how many do you think I will get?"

Helen shrugged.

"Probably 25 percent as a guess. Can you actually do that?"

"Sure," Carl smiled, "this new software program lets me sort the list anyway I want."

"You might consider selecting the ones who have had two, three, and more problems with us."

"Doing," said Carl. "Is that your phone?"

Helen scrambled around in her purse, snatched out the cellular unit, and answered. She responded in the affirmative several times and hung up, smiling.

"That was my bookstore," she reported, "they figure I am not going to make it before they close so the lady offered to drop the book off at my home on her way. How's that for a complete transaction?"

"It would be a good example for our suppliers. We have roughly one-third who have given us zero problems, one-third who have screwed up once or twice, and one-third who are always in trouble. Sounds like we could make some quick progress this way. That 30 percent must take up most of the time our purchasing people spend on corrective action."

He turned away from the screen and extended his hand to Helen.

"George asked me to find someone who would like to take on the job of managing supplier relationships. How would you feel about that? It would be a promotion."

Helen thought for a fraction of a second and nodded enthusiastically. She was ready for a new challenge and this was just what she had been looking for.

"I'd like that, Carl," she smiled, "however, I would like to concentrate on the Complete Transaction idea. That could include doing a study to expand that into a performance assessment method for our employees. If it works out, our suppliers and employees, who are also suppliers, can be evaluated on the same scale. The purpose would be to find out who wants help and provide it to them. Then we would be on our way to having veteran employees and suppliers."

"Sounds good to me, but you'll have to talk to George about it all. I'll tell him that you are my choice for this supplier situation and you can talk him into the rest. Give me a chance to chat with him in the morning and then go see him. Good luck on this. You might be able to get the whole world to be complete. Perhaps our local ball club might even be able to win a game now and then with your help. After all, they are a supplier too, of sorts."

A year later Helen made a presentation at the monthly management meeting concerning the status of CTR. She had coordinated her report with the Purchasing, Human Resources, and Quality Management departments. She explained how they had all worked together to incorporate CTR as a routine part of the way the company worked.

"I would like to begin by saying that during the past year with the introduction of Complete Transaction Rating we have reduced our supplier base from 2500 to 675; we have reduced our purchasing costs by one-third; we have seen problems with quality in purchased products and services virtually eliminated; we have had a very positive employee participation in the personnel evaluation aspect; and we have reduced customer complaints to a trickle. We are now locked in deadly combat to eliminate that trickle. All of this has come about without the addition of a single new employee or the expenditure of any capital equipment money. We have had to develop a new software program and conduct some management training.

"It is easy to say that business consists of transactions and relationships, but saying does not necessarily graft the understanding to us. The whole key to management is that we ask our employees and suppliers to perform certain transactions for us. The result of all these as a whole is the company's structure and its output.

"All the improvement programs and philosophies we have examined deal with communication between employees, management, suppliers, and customers. However they really don't get specific about it. Specific is when everyone knows exactly what everyone else is expected to accomplish and what the result will be if it is not accomplished properly. This way each person and supplier knows the importance of their role. No one is going to check up on them until the result of the transaction is applied.

"When we think in terms of transactions, we then have to have clear requirements that describe what must happen for it to be complete; we have to have clear training so the individuals can understand and relate to them; we have to have a clear way to recognize the possibility and need of improvement. With this way of doing things we actually describe the operation of the company in tiny detail so easily that everyone is happy instead of complaining about being bound by procedures. In most cases, the description of the transaction is prepared by the individual who is going to do it. Whether or not the result is accomplished completely is easily determined by the one who receives the result.

"Incidentally, I do not want to give you the vision that we have immersed the company in piles of paperwork and procedures. Actually we have learned to use as a sort of E-mail experience a computer program. Someone wanting to send a package to London, for instance, can call up that transaction on their screen and determine what we have learned about how to do that. So we build experience files, and improve them as we learn. New employees are shown, during their orientation and training, what we know about the transactions they will perform.

"Let me give you two examples:

"Suppliers. We had never made it clear to suppliers that we expected them to provide us with exactly what we ordered. To make that happen there is a series of transactions on our part: first we had to determine what we needed, a software program for instance. Second, we had to describe, perhaps with the supplier's assistance, what we wanted the program, product, or service to accomplish. This is placed in the computer system and the supplier is given access to that part. Then a constant communication can take place until the content is agreed. At that time, the supplier begins the process to deliver what we have ordered.

"When we began this approach we notified all of our suppliers that we would not accept anything that did not conform exactly to the requirements agreed with the purchase order. We also told them that we were only going to have one supplier for each product or service we purchased. About a third of them dropped off the screen right then. They never responded to our message. We are developing a corps of sound relationships with those who did respond.

"We rate suppliers according to their real-life performance. Price, schedule, and quality all have equal ratings on the CTR. We learned that going around and doing surveys proved nothing. Everyone can look great for a day or so. And things change rapidly: a new executive or a failed proposal, and the supplying company is a different place. So we deal only in performance, and that continues for the life of the supplied item. At one time everything a company received was checked out at the dock; if it passed that examination, the supplier was home free. Anything that occurred later on was assumed to have been caused by the purchaser. However now we look at everything throughout our whole process as it is used, right into the customer's shop when practical. Computer entries can be made anytime. This gives us a wealth of information about the supplier and their reliability. It is a big help to corrective action efforts.

"Personnel. As you know evaluating employees has always been difficult because so much subjective information is required. I know that I have been concerned about receiving and giving personnel evaluations. People are not always given accurate or fair ratings just because there never has been a system developed that stuck to the facts. However, with the CTR concept we are able to deal directly with employee performance in the same manner as suppliers. If we think about it, employees are suppliers also, and this gives us a way to look at actual results.

"Each of us has customers inside the organization, just as we all have internal and external suppliers. We have to determine what our customers need from us and what we need from our suppliers. If I am supposed to provide the weekly sales report, for instance, I need to understand the format expected and the content. Then I need to know where to collect the data that will make a complete transaction. When my report is turned in on time, containing the correct information, then the transaction receives a 1.0 rating. Since the report is transmitted on the computer it is automatically logged as to transmission time. Those who receive it can note whether it is complete or not. I can check this result to learn if I need to take corrective action with myself or my suppliers. Over a period of time a pattern of results develops. If the report is on time, every time, that is great. If it is late 30 percent of the time or something, then we can see that I may need some help.

"The purpose of the CTR is to help the employee, and the company work flow, to be accurate, efficient, and consistent. It also lets management determine who their best workers are. Under the old systems a lot of evaluation became popularity contests. The best workers may not fit into that category.

"What we have learned so far is that the employees really like this way of having real-time evaluation. It is possible to check the results of your output whenever you want. Customer complaints are registered as they occur, for instance. I can inquire from that screen in the corner and get a live report of what is happening today as well as historical data. There is no reason for anyone to be out of date."

One of the executives raised his hand.

"Are we creating a 'big brother' environment here, Helen? Do people feel we are spying on them?"

Helen looked puzzled for a moment.

"The information system is open to everyone, to make inputs and collect data. If anyone feels at anytime that something is incorrect, they only have to enter that on their screen in order to have it reevaluated, or they can deal directly with me. We get very few of those, and we take them very seriously. That has led people to have confidence in the integrity of the system.

"The part that may be difficult to accept is that companies have never let the employees participate in the minute-by-minute running of the place. Now they are active and we get transaction improvements in a continual stream. Our productivity is increasing dramatically."

She turned to the Operations executive.

"Would you care to comment on that, Otis?"

Otis smiled and agreed.

"There is an openness about operations now that I have never seen. People know that they will be recognized for the work they do. Before, this was up to their supervisor and that brought up a lot of personal problems. Now it is there for all the world to see. Productivity is up about 20 percent and we haven't changed anything. I must say that Helen was very careful to make certain everyone understood what was happening and what it meant to them. She has taken away all the 'big brother' thoughts.

"However," he continued, "we have to be vigilant and make certain that we do not let someone turn this into a negative system. For that reason, I have suggested that we form a revolving group representing all areas of the company to review the usage continually. CTR is a great competitive tool, I would not want us to abuse it."

The CEO thanked Helen for her comments and asked that she think about forming the suggested group. A different member of the executive committee should be on it each period, he thought. She was asked to return each quarter and give the executive committee a status report and was invited to return any other time if she had anything she wanted to say.

RELATIONSHIPS

Relationships with employees, suppliers, and customers are the other side of the business equation and their success can be measured also. As in personal matters, relationships are a function of paying attention to each other. But it cannot be contrived attention; it has to be real, and useful.

Consider the employees, who far outnumber management and in large organizations may never actually lay eyes on the really senior executives. Relationships with them lie not so much in personal contact but in the way they are treated. There is a marine general whom I never met that I feel very close to. He made a policy that each of us would get one hot meal a day regardless of what was going on. This took a bit of doing when combat conditions prevailed or we were moving an extended distance. But that policy showed his concern for his troops far beyond strong speeches or standing by the side of the road waving as we struggled by. He put his resources where our mouths were.

The same is true in business. When workers, whether in an office or a factory, feel that they are just a cog, they act like cogs. Letting small groups do complete jobs is one way of getting around that feeling. Having formal empowerment programs is a way of showing employees that management does not care about them. It is like sending your kids a catalog at Christmas instead of going out personally and selecting something for them. Catalogs aren't personal. People need the opportunity to communicate, they need to receive information. There is nothing complicated about all this. I would be careful to make certain that those who manage such things in the organization are sensitive people. Don't pick the one who always forgets their spouse's birthday, or never asks others to share something they have.

Successful supplier relationships are based on straightforward policies concerning performance and the opportunity to communicate. Regular seminars for suppliers and the purchasing people will bring an understanding of the need for coprosperity.

Successful customer relationships come from performance and reaction. Everyone knows this but very few do anything about it until a large problem emerges. The company must have clear policies on dealing with customers and the training to implement it. Leaving such things to the natural goodwill of the individual is not very reliable.

Management has to review its status on relationships continually with the same intensity it reserves for the financial reports. Companies can survive financial problems and even quality problems. Relationships are fatal when they are negative. For this reason, executives have to really understand that relationships cannot be left to chance or the flow of business. "Attention must be paid," as Arthur Miller wrote.

HOW CONCEPTS ORIGINATE

When someone offers you their view of the truth you have the right, and the responsibility, to ask them to tell you how they came to those conclusions. Are we dealing with the opinions of a lecturer who has been listening to himself all these years? Are these truths proven scientifically? Sir Francis Bacon talked about ants, spiders, and honeybees in this regard.

Ants collect information, process it, store it, and retrieve it. They do not contribute anything to it, and they probably do not understand it.

Spiders develop information from their own mind and spin it out in a web that is very likely looking. Yet they have little real experience in the things they put out and they are not interested in what others have proven through controlled experiments.

Honeybees gather information from everywhere, process it to make certain exactly what it is, and then arrange it into something proven and useful.

I hope I come to you as a honeybee. Everything I state about managing quality and business is something I have experienced or measured. I would like to make certain that the reader knows from whence they came. It is important to those who would be successful.

When I talk about management systems being as real as the tooth fairy, the Easter Bunny, and Santa Claus, people naturally want to know "so what then should they use to manage?" The answer, of course, is concepts as opposed to techniques and canned procedures. We run our lives on concepts; no computer program can handle a marriage, nor can a set of processes deal with our budget. We have to have things we understand and believe in to serve as the foundation of all our actions. We learn early in life that it is folly to try to breathe under water; we are not designed that way. No matter how liberal, conservative, stubborn, or even stupid we are, this concept stays firmly in our minds. We learn that if we spend more money than we earn, we will be in trouble. Knowing does not always keep us from doing, but at least we know where the problem lies. Just because something is true does not mean that people will embrace it. Those of us who have had automobile accidents realize that we usually broke some basic commonsense concept in the process; usually we had been warned not to do exactly what we did. However, good advice doesn't by itself generate good management. Lists of 14 aphorisms wind up being entertainment; they don't get into the fiber of our management drive. My friend Bob Vincent used to say that "beauty is skin deep, but ugly goes clear to the bone." It is sort of like that.

We can't hope to memorize and apply everything we hear or see but we can lock the critical few into our heads when it comes to managing something. Concepts are truths that apply to all situations and serve as a basis for evaluation. In quality management these are the Absolutes:

Quality means conformance to requirements.

Quality is obtained through prevention.

Quality has a performance standard of zero defects.

Quality is measured by the price of nonconformance.

This chapter is aimed at discussing the personal experiences that led me to isolate and identify these concepts of quality management that have worked all these years. It is also a lead-in to a description and discussion of the PCA case where these concepts, plus the idea of transactions and relationships, were used to deliberately create and manage what turned out to be a very successful company. We are talking about the real world here, and I know I keep saying that, but we have to understand the difference between guesswork, false dogma, and reality. It is not always easy to separate them. Anything worthwhile in business is measurable; opinions are interesting, but real things must stand the test of measurement. When I tested the conventional wisdom of quality that was preached to me, I found that it was not in touch with reality. That is when I began looking for real life.

Let me go quickly through the learning events that caused me to come to these understandings. I served in World War II and joined the Naval Reserve at the time of my discharge. This was not one of my brighter decisions, as I was recalled to the Navy for the Korean War after graduating from the Ohio College of Podiatric Medicine in 1950. When I was released from the Navy in 1952 I was too broke to think of practicing medicine, so I went to work at Crosley in Richmond, Indiana, as a test technician in their quality department. This was my first exposure to business life. I enjoyed it so much that I didn't even realize that I was on the lowest level of the organization. My education was medical and my experience in the Navy was concentrated on holding sick call and treating field wounds suffered by Marines. I had no experience with electricity, electronics, or machined parts. All of those items were involved with my work in the assembly area. People were patient with me but it was obvious that they regarded me as someone who was well intended but uninformed. As I began to understand what was happening in the processes, I often had ideas which I felt might prevent some problems we were having. These ideas were greeted with the tolerant smile one presents to a child—and ignored.

Then one day the engineering supervisor, who was experimenting with a new process, was cut on the wrist by a whirling antenna; blood flew everywhere. The whole place froze and then panicked. People were yelling and screaming while they ran around. But this was what I had been dealing with through two wars. I grabbed him and laid him on top of a workbench, shut off the blood eruption, asked someone to call for an ambulance, calmed him, and put a homemade blood-flow block on his arm. He did not go into shock since he realized quickly that everything would be all right. The ambulance came, took him away, and we went back to the lines. He returned to work in a few days. All of this meant little to me, as I had been through it many times. But it made an incredible change in the way I was treated by people in the company. Now they thought I was smart and useful.

Apparently being found competent in one thing gives you some credibility for being competent in other areas. Suddenly my ideas for process improvement or problem prevention were taken seriously and often, to my surprise, actually worked. I began to think that perhaps I could carve out a career in this unfamiliar world of business. It was becoming apparent to me that the way things were done was not very efficient. A group would assemble something, some others would inspect and test it, and another group would repair the inadequacies. The concepts of quality control were used to run things. These ideas stated that errors were inevitable and that it was necessary to plan for them. This mindset led to the formal establishment of "Acceptable Quality Levels" which was a kind of comfort zone concerning the amount of defects. All of this conflicted with what I had learned in my medical education and experience. The man who had been wounded by the antenna represented only one case in thousands of opportunities for failure. We should have just let him bleed to death rather than having all that lost time since he was not statistically significant. That is how silly all of this sounded to me.

I had joined the American Society for Quality Control in order to learn more. The local section was very active and that was where all my friends were. One night we had a well-known expert who was kind enough to come speak to us. I asked him after his speech, when no one was around, why all the effort of the profession was directed toward justifying doing things wrong. He said that we had to deal with the real world. I said that I thought we were creating a false world with the concepts of quality control. He placed his hand on my shoulder and told me, kindly, that I would understand after a while. At that moment I realized that I was talking to the tooth fairy. Life is a lot easier if you deal in mythology. I went back convinced that things would work out better if there were some way to turn the world on to prevention rather than detection.

At this time I joined Toastmasters to learn how to speak and began to think about carving a career out of the quality business. It was a highly technical area, and I was not a technical person; it dealt with statistics which bored me. I would never relate to those activities but it seemed to me that there was a great deal of room for managing quality, for making it happen rather than just dealing with what floated down the stream. It also began to dawn on me that an assembly line was not the platform to transform business philosophy. I had to get out of there.

The Richmond plant was a branch of the main operation in Cincinnati and the local staff was mainly workers and a few managers. Once in a while executives would visit us from headquarters. They all dressed and spoke well, and as far as I could see, got to do whatever they wanted to do. In my area we had five minutes an hour to visit the bathroom, grab a smoke, or do whatever could be done in that time. We had 30 minutes for lunch and received a reprimand of one sort or another if we overstayed our leaves. Once I lingered before going to lunch in order to supply some information the department manager had come to the area to obtain. I added that linger to my lunch time and was given a lecture about it. I realized for sure that respect came to people in business according to their job level. When I became an executive, I promised, everyone would be treated equally. No one had ever mentioned such a thing to our managers and executive. So I decided to do what was necessary in order to become an executive. This would require an opportunity, and my preparation. The key would be helping, I would help people succeed.

Since we were slowly starving to death in Richmond (making $315 a month), and there was no sign of a career path, I searched around and found a job in Mishawaka, Indiana, at the Bendix Corporation. This was a plant where they were making the Navy's TALOS missile, a ramjet surface-to-air vehicle to be launched from a ship. Here I became a reliability engineer responsible for the assembly area. They all believed the same quality-control things, and could prove it by the reliability statistics. I had learned to keep my radical ideas to myself while plotting to overthrow the government of quality.

At Bendix I was free to get out in the area and help the manufacturing and engineering people produce missiles that the customers could actually use. TALOS was still in its development phase. My main work was to look at the nonconformances that the inspectors and testers had identified and then classify them as to seriousness, cause, and responsibility. This was placed on a form which was given over to the IBM center. Now we are talking 1956 here and IBM was in the punched card business. Computers were just beginning to edge into the business world and one the size of a three-bedroom house had the capacity of something in a wristwatch today. The IBM center produced printouts of all the nonconformances and charts were made from those printouts to show the quality status. Acceptable Quality Levels (AQLs) were assigned to each area and as long as the line on the chart didn't go lower than that, everything was going just fine. One day I determined that a nonconformance was caused by an engineering error. So I went up the stairs to that area and found the group engineer for that particular system. He was very interested in my defect form and the fact that I existed at all. After much conversation it was revealed to me that the Engineering Department had no idea that all these nonconformances were classified and reported. They had heard of the Reliability Department but had no idea of what they did. They didn't realize that the assembly area was struggling each day trying to put these missiles together when the design was not all that clear, and changed continually. I went down to the IBM center and swiped one of the closely guarded printouts, which weighed at least 25 pounds, and took it up to my new friend. Soon he and his fellow group engineers had gathered and began to pore over the data. The result of this activity was the assignment of engineers from each group to respond to my calls for action. This idea was their own; they thought it might be possible to find some current problems and use that information to prevent future situations. As part of this effort I introduced them to the Manufacturing management so they could open a dialogue, which they did.

Armed with this success I marched over to the Purchasing Department. They considered themselves to be in the service business rather than manufacturing. That is what the department head told me right off. They just ordered whatever was sent up to them to order, they got an AQL from the Quality-Control Department when necessary, and outside of that it wasn't their problem if something didn't work right. They had a system for sending things back to suppliers, but to his knowledge nothing had been sent back in months. At my request he made a call to confirm that this was indeed the case. I pointed out that according to the printout I carried up to his office, there were hundreds of nonconformances charged to the suppliers. Didn't he feel some responsibility for all this? He replied by asking me if I didn't have something else to do. I left the printout on his desk and went off to find out what happened to the supplied items that did not conform to the company's requirements. It turned out that we had an extensive and competent repair team that could fix about anything. So when something was rejected it just went to that room. This was part of the formal Material Review Board which could dispose of anything as long as the paperwork was proper. The supplier never knew anything was wrong and kept doing it that way, and purchasing never heard about it. We were dealing with the Easter Bunny here. All of this let me realize that this whole system had developed because requirements and specifications were not assumed to have real meaning. Everything was made to be used in some fashion. It took a lot of work, and people were proud of being resourceful. Every missile was different; there was no learning process, just survival. The concepts that error is inevitable and that we have to make do with what we get were making it impossible to have a smooth running operation. This was when I began to realize that we had to establish exactly what the word "quality" meant if we were to get things done right. We had to have a concept of "conformance to requirements" if we were going to learn how to have a smooth, efficient work flow. If everyone got to make their own definition, then our world had no substance. But everyone was satisfied; we were certified to Mil-Q-9858 and approved by the government. The fact that not much worked right had little to do with it.

One day my supervisor got promoted. His office was in a little glass box and the several reliability engineers had their desks in a line outside his box. The engineer in the first desk moved into the office and the rest of us all went to the next desk. I was now three desks away from being the supervisor. Obviously, it was time for me to get moving to a place where there was a future. I began to look around and on a trip to Boston in the middle of February 1957 I saw an ad in the local paper. A new organization was forming in Orlando, Florida, and they were looking for reliability engineers. Their people were interviewing in my hotel. This is one time my habit of reading everything paid off. They made me an offer that evening, contingent on meeting my future boss in Chicago the following month. In May my family and I packed up and moved to Orlando to join the Martin company as a senior quality engineer making $640 per month.

Martin grew to 10,000 people over the next eight years and I had a marvelous opportunity to learn. As a department manager a few years later, I was in charge of the Pershing quality program which involved a thousand people in the quality and testing areas. It was during this time that I developed the concept of Zero Defects (ZD) because we could just not learn how to find everything that could be wrong in a weapons system. We had to prevent, not sort. As I have written before, ZD was taken by the government and much of the aerospace industry as a motivation program rather than a management performance standard. The powers of quality control thundered down on me, but I could see the reality of what was happening on Pershing. Later, when I was managing supplier quality, we were able to help the suppliers get us the right stuff, on time.

But the most important thing I learned at Martin had to do with relationships. When the company was building its employment rapidly I was concerned with how we were going to let all these new people know about quality. Most of them were local and had no business or industrial experience. Some thought their lack of experience was a problem, but I looked at it as an advantage. Later in my career I searched deliberately for people with no experience on the thought that it was easier to teach them the correct way than for them to unlearn their bad habits.

My car pool friend, who worked for Personnel, told me that they were beginning an orientation program for new employees. He wondered if the Quality function wanted to do something along that line. I asked the director about it, and discussed it with a few of the managers, but they thought it should wait until people started actually working. I didn't agree and so I went ahead and started talking with each new group as they joined the company. It only takes 15 minutes to explain that happy and successful work comes from learning how to do the job right the first time. The Security Department spent eight hours on its part of the orientation program, all of which was quite dull. My few moments were a bright spot in their day. The effect of all this was to lay a base of quality awareness for employees that paid off as they went to work in accounting, manufacturing, marketing, engineering, maintenance, and the dozens of areas in this complex facility. People who wrote the instruction books for the customer, for instance, realized that they had to understand what those in Engineering really wanted to happen. An unplanned side effect of the orientation course was that most of the people in the company recognized me on sight later. Everywhere I wandered in the facility people would speak to me and talk about quality. That helped them be successful.

When the Orlando operation was just a couple of years old it got into trouble with the customer on quality. The general manager had decided to reduce costs so he cut out inspectors in the shops and gave the acceptance stamps to manufacturing. All of this was done without training or control. Inspectors are no more ethical or honest than production workers, but they are trained and supervised concerning their profession. It wasn't long before nonconforming material was all over, and out of, the place. The general manager and the quality director were both removed and Tom Willey, an old-time Martin executive, came in to run the establishment. He appointed a new quality director, Jim Dunlop, Jr., who had no experience in that field but was a successful project manager. Jim asked me to help him restructure the quality operation. This was a wonderful opportunity to implant prevention in a large organization.

Mr. Willey spent his time in the plant with the people, finding out what was happening and building morale. As part of this he started inviting employees, about 400 at time, to bring their spouses in to see where they worked. We would have an evening program with tours, dinner, and some remarks. Everyone was treated with equal dignity. Many of the executives were not comfortable dealing this closely with their employees. It was at this time that I began to realize that there was progress to be made in my career by being the interpreter between the employees and the executives. The bigger their offices, and the broader their privileges, the less they knew about reality. My mental agenda of progressing toward becoming an executive focused on the inadequacies that seemed to go along with the job. For some reason when people reached that position of authority they became antagonistic to the levels they had previously occupied. They placed people in categories and assigned bad characteristics to them. One heard similar statements in executive suites all over:

"People don't really want to work hard."

"They don't care about their work."

"All they think about is money."

"Don't buy a car made on Monday or Friday—the workers are really bad then."

"You'll have to explain that carefully or they won't understand it."

The workers had their own list of clichés based on management's reluctance to listen or be considerate. These thoughtless put-downs separated the two groups. I made up my mind that I would never take either view. People would stand or fall with me based on what they were and did. The concept of treating everyone the same is valuable. It is the best defense against becoming arrogant, which is the fatal disease of executives. I said this during a speech in India and was surprised to see the audience, which consisted of senior level people, agreeing enthusiastically. Usually such a comment is greeted with frowns. Then I discovered that in India the word "executive" refers to the level we consider to be "manager" in the Western world. Their top level is called "manager." Keeps you humble, that stuff.

Jim Halpin became the quality director when Jim Dunlop wanted to go back to project management. We made Dunlop an honorary quality engineer. He had done a good job. Halpin was an administrative genius to my eyes. He really knew how to get things done and had the complete trust of Tom Willey. We went to work on the customers to show them that we had learned the error of our ways and were running our organization properly. Jim was very patient with me and taught me to keep my jacket on, not wear white socks, be concise, and let me innovate. When the Pershing program became our biggest project it had been divided into three separate departments: Ground, Air, and Field Test. I was running the Ground Equipment Department, as the junior manager in the division. When all the departments were combined into one project, he selected me to be the senior manager. This caused a lot of consternation among the other managers, but following Jim's advice I met with each of the managers personally, in their office, and asked pleasantly if this was going to be a problem for them. All assured me that it would not be and I took them at their word. The Pershing Quality Department involved over a thousand people so we didn't have much time to be political. After a while they realized that I knew what I was doing and there really was no problem.

As we moved along developing the missile system and beginning the firing program at Cape Canaveral, output became slower and slower. The bottleneck was final testing of the missile itself which was conducted by engineering personnel. The strategy was not to turn it over to the hourly paid people because that would bring union situations which would disrupt work. For this reason test engineers and technicians were conducting system tests which took two or three weeks to complete. I suggested several times that we had no disruptions in other areas where the lower-level testers worked, and that the engineers kept reinventing the systems all day. This comment was not well received by senior management. It seemed to me that I should be considered senior and be invited to important meetings since I had a large department and was dealing with the customer all the time. However, quality was considered something that we were obligated to do by most executives.

One day Tom Willey went down to the test area and tried to find out what the bottleneck was. He could not get what he considered straight answers, so he went back to his office and made a few telephone calls. The result was that final test was transferred over to my department, which was the only one that had not asked for it. Who needed all those people and their problems? I moved all the test engineers off the assembly floor and up on the balcony. We brought hourly testers into the area and made certain that the test procedures were up-to-date. Then I instructed the supervision to run tests until they got a red light. At that point we would determine if the missile itself was incorrect, if the test equipment had a problem, or if the procedure was not right. We worked out each of these red lights properly until it became green, then we moved to the next red light. This took a while and people became very impatient, particularly with me. But this was becoming another concept development, there was no advantage in learning how to do the wrong thing quickly. Slow and steady wins the race is still pretty good advice. Tom Willey and Jim Halpin stood by my strategy, and all that fussing didn't bother me. The Marine Corps tends to teach focusing; it doesn't do it officially, it just does it. After a while we were able to test a missile and confirm that all its electronic systems were functioning properly in 45 minutes. Soon the place was running smoothly, the engineers had begun to cooperate willingly, and everyone was properly amazed that the hourly paid testers had turned out to be responsible and reliable. The problem here was that those who were concerned about the lower-level workers had never actually worked at that organizational level. It was a foreign country to them.

However, Manufacturing still had its own way of looking at things. One day they performed some planned modifications on the telemetering equipment before it followed a missile to the Cape. This confused the integrity of the acceptance testing. I called the missile and the equipment back, even though it was speeding along Route 50 toward Canaveral. This delayed everything and caused a lot of blue words to sail through the air. We retested the system, made the corrections, and sent it off a day late. Those who had assembled at the Cape for the flight test just had to spend another day at the Surf Club. The effect of this exercise was to make the point about taking quality seriously once and for all. We never had a problem like that again. Tom and Jim protected me from the lynch mob.

I have written about creating the Zero Defects concept in *Quality Without Tears* and other books so it is not necessary to go into it here. But that incident was the kind of thing that made ZD necessary. Those Manufacturing folks who updated the telemetry equipment to where it no longer matched the missile had never paused to think about the effect of their act. They were just doing their job as they saw it. The purchasing agents who ordered materials and services were not conditioned to consider that someone had to actually use the stuff. Working with that department we began to bring suppliers into the plant in order to turn them on to getting things done correctly. These seminars were conducted on a continual basis since we had a great many suppliers. When I took over Supplier Quality Management in 1963 we made it a goal to eliminate the need for receiving inspection and test. This never quite came to complete fruition but we did reduce rejections to a small number and returned almost 100 percent of everything that didn't meet its requirements. Supplier quality went from a big problem to a small problem in just a year. This came about when the concept of Zero Defects became an actual working policy for the Purchasing and Quality management.

Around this time, I had a discussion with the senior people in Personnel concerning my future. They were pleased with my performance, and I had been on a quick promotion track; however, all that was coming to a close. Since I was not an engineer and, in fact, did not have a very useful education, there was no probability that I would go to corporate headquarters one day. My future was going to be as a department head. I attribute this to the conventional way of looking at things in big companies. You are known, always, as the level for which you hired into the company. I was considered a "bright young quality engineer" even though I had progressed far past that point. For me to become the director, or go to headquarters, was unthinkable. This mental block became another of the things I said would never happen in any organization I ran.

As a result of this rejection, I began looking around and in a year stumbled across the ITT Corporation in New York. They were looking for a quality director, and the CEO of a supplier company recommended me. ITT was mostly in the commercial field and operated overseas. Harold Geneen was the chairman and CEO and he was building the world's first conglomerate. Using real money, he eventually purchased or merged with 500 companies. The company was not quite $2 billion in revenue when I joined in 1965, but was $20 billion when I left 14 years later. This was an exciting company and an exciting opportunity. Their concern about me was involved with two things, which I discovered several years later: first, Why was Martin not asking me to stay by giving me a better job? and second, Since I came out of the defense industry would I be too strict for the commercial world? Both of these issues were resolved and I went to New York in May 1965. My family and I found a home we could just barely afford in Greenwich, Connecticut, and moved. The prime family memory of that trip was passing our church in Maitland on the way north and hearing six-year-old Phylis say, "good-bye God, we're going to Connecticut."

ITT was in a lot of businesses: telecommunications, defense, rent-a-car, semiconductors, baking. It added automobile components, insurance, hotels, forestry products, financial operations, frozen food, and many others. It became apparent to me quickly that I was never going to understand every one of these businesses. It was also apparent that installing a company quality management specification along the lines of Mil-Q-9858 would be as ineffective as the Department of Defense found it. What I had to do was create an organizational culture that understood and practiced the concepts of quality management as I had conceived them over the years. There didn't seem to be anyone against quality; in fact, all the senior people I met were apparently looking for me to cause a few miracles. As I visited a few operations and looked around, I felt that their output was as good as others in their fields. They were mostly satisfied with it and the customers were adjusted along that line. This was an early example of what is called "Benchmarking" today. It didn't relate to how effective and productive it was possible to be. We didn't have many big problems, just what was accepted as being usual in that day. There was a continual struggle to get problems found and resolved; every unit had a list of "quality problems." My boss thought I should be out solving all of them; I thought we needed to prevent them from happening. Since I was the only member of the corporate quality function, it was impossible to be in all the operations at the same time, so he reluctantly let me do what I was going to do anyway.

I searched organizational charts for the quality professionals in the manufacturing operations; there were none in the service unity. Meeting and chatting with these folks showed that they were all imbued with the conventional wisdom of detection. They were nervous that I was going to insist that everything get done according to what had been agreed. They were right. There was a great deal of money involved in doing all these things incorrectly and being so good at problem solving.

To attack the situation, I decided to concentrate on policy and education. First, we had to create the playing field by establishing clear policies and practices; second, the quality professionals, the corporate executives, and the operating management needed to be educated. They had to understand their personal role in causing this culture we needed. Most of the people I met were interested in having me get the other people on the right track.

ITT—SETTING UP TO MAKE QUALITY CERTAIN

While thumbing through the policy and procedures manuals (one of which had 85 pages on how to properly use the ITT logo), I discovered a notebook of organization charts. They included all of the subsidiary companies worldwide, complete with a listing of their managements' names along with the addresses and telephone numbers of all the facilities. There was no mention of any Quality Department or person in the package. However, there was enough information to let me conduct a telephone tour of the U.S. companies in order to find my quality managers. To a man, and they were all men, they were delighted that I was there in headquarters wanting to be of help to them. To a man they politely let me know that it was not possible to rescue them or get through to their bosses. None of them reported anywhere near the top of their organization, except a few in the defense group. None I talked to had any idea of how their business was doing financially. They had plenty of problems and were not sure that they would be permitted to come to a meeting in New York. The common thought was that it would take them away from the operation too long.

After this experience, I examined the corporate policy and procedure books in more detail, searching for something to build on. The company needed an original and concerted quality effort beginning with a policy on quality. It would have to be a corporate statement that would eliminate arguments and confrontations up front. I wrote one:

> The quality policy of the ITT Corporation is that we will deliver products and services to our customers and co-workers that meet the agreed requirements, or we will have the requirements officially changed to what we and the customer really need.

Then I added a couple of paragraphs that required every unit to have a quality management function reporting to its senior executive and performing tasks as agreed by the senior executive and the corporate quality director (me). I also included a policy goal that we would "make ITT the standard for quality, worldwide." The Policy and Procedure Department was on the next floor so I took my notes down to them and asked that they be issued. In the process I learned that New York office buildings had limited access from the stairwell. You could get into the stairwell but the doors didn't open onto a floor. I had to walk down to the lobby and take the elevator back up to eight to reach the policy office. It is things like this that keep you humble.

The policy people were most cooperative; they noted that I could sign these new policies but Harold (Hal) Geneen, as CEO, had the final approval. They would send the final forms up to him on the twelfth floor, unless I wanted to carry them personally. I thought we would just let them do it, they did, and he agreed right away, asking that I sign for the corporation. This was a good strategy. Later I discovered that this quick response was unusual—his paperwork usually was in a time gap of a week or so.

I didn't expect a revolution from these policies but it laid a foundation. I made certain that copies were sent to each quality manager worldwide, as well as their bosses. Most of them would have never seen it otherwise. It is interesting that companies have a limited distribution of their policies, probably because they are typically placed in expensive leather notebooks.

Writing something for consumption by others in this big world headquarters was nowhere near as easy as in my former life. Secretarial help was scarce and everything took a while. I noticed an old IBM electric typewriter back in one of the storerooms, and after a halfhearted search for its owner, I moved it to my desk. Now I could write my own memos and letters. I lugged that machine from office to office all my years of working there. Anyone who tried to read my handwriting was grateful for its printed letters. As the only executive who typed, I was considered a little weird but that is an advantage in an organization where people strived to not be different. When I did write some memos and trip reports, I found to my amazement that they were actually read by those who received them. In the world of Aerospace it was considered bad form to send memos out to wide distribution. In ITT they complained if everyone didn't know what someone had written. Each day, for instance, I received a pile of pink telex copies, all the ones that were sent by anyone in the system, at least 90 of them. When I tried to get off the distribution list, the folks I talked to acted like I was becoming a Communist or something. Finally I just tossed them out each day if they were not addressed to me personally.

Looking through the paper clip closet, where pads and pencils were kept, I discovered a little booklet written by the CEO. It laid out a way of writing a memo and was titled "Unshakable Facts." It said that the findings of a business trip should be listed first, one through five for instance. (The example I use to explain it is: 1. Burn the plant down. 2. Fire all the management. 3. Plow the ground with salt.) Then a paragraph should be used to provide more detail on each of these recommendations. This should state the reason for the suggestion. (1. Nothing profitable has ever been made here.) Lastly the author's unvarnished opinion should be stated. The findings and recommendations can be fought about and even rejected, but the opinion cannot be altered. Right or wrong, the writer is entitled to say what he or she thinks.

I immediately began to follow this layout and wrote all my reports that way for the rest of my career in ITT. No one else ever did, as far as I know. That always made me wonder. When the big boss lays out a report format that makes sense, why not use it? When I finally got a staff, I had a hard time getting them to do it even though they liked to read my reports. After a few years I finally gave up trying to change people. (One day I will write a book about how fruitless it is to do management development programs.)

My first trip as the ITT quality director was to a plant in the midwest of the United States. Before venturing out of headquarters I had tried to get a feeling about the relationships between the units (as the subsidiary companies were called) and the headquarters staff. I knew that in Martin everyone hated the Baltimore corporate people. This may or may not have had basis in the fact of their actions; I hardly knew any of them. In ITT the staffs had the reputation of being very tough on the unit people. The old story, which I stole somewhere and have used for years, is that they are called seagulls: "they fly in, squawk a lot, eat your food, crap all over you, and then fly away." It seemed to me that a headquarters "expert" who was useful and didn't make trouble would stand out in that environment.

I made certain that I started my visit in the plant with the general manager, whom I had called ahead of time to make an appointment. As part of the visit I rejected their offer to pick me up at the airport and made my way over. My feeling was that we could have a stronger relationship if I acted like I was working for them; there would be no special treatment. The general manager was very gracious and as we sat for a while I asked if there was anything I could do to help him. What was his biggest problem? After some reluctance he decided to share with me that there had been a sudden burst of customer complaints about receiving products that were different from those they ordered. I said I would look into it. We talked about the role I planned to play in the corporation and wondered if it were possible to influence a worldwide conglomerate in something like quality. After a while he called the quality manager and asked him to come to the office to pick me up.

I spent the morning getting familiar with the operation. They ran a typical machining and assembly shop. There were a few statistical process control charts, but basically they just shaped up the little ones and then made big ones out of them. The products were well designed, were proven over the years, and were considered very reliable. The quality manager, Harry Walters, got comfortable with me after a while and began to chat about his problems. There was a very strong marketing group and they were always pressuring the shops to get one thing or another out ahead of the plan. This led to a lot of unnecessary confusion. Also, Marketing was always changing configurations in order to appear to be offering customers the latest technology.

After a while a call came down from the general manager asking me to come up to his office for lunch. Harry delivered me there and I was led to the executive dining room where the directors of Marketing, Manufacturing, and Finance were waiting. We all introduced ourselves and sat chatting while the meal was starting to be served. I looked around the room and when the marketing director asked me what I was looking for, I said I was wondering where Harry was. He didn't belong to the executive dining room, I was told, but they could invite him up. Acting on this afterthought, they made a couple of calls and soon Harry arrived rolling down his sleeves and very nervous about the whole thing.

I chatted up a storm while eating lunch as the rest of them poked their food around. It was in situations like this that I learned what was really happening in the corporation and in this particular industry. Finally the subject of quality came up and I led the conversation around to the problem of the wrong products going out to the customers. The marketing manager launched into a condemnation of the manufacturing operation, which was immediately refuted by their director sitting across the table. After a while I asked Harry what he thought caused the problem. He shrugged his shoulders and said that he had included his thoughts in his last report; the problem was that marketing had changed the sales catalog but had not told manufacturing about it. The result was they were shipping the right part number but it related to the wrong part. He had been unable to stop it or even get anyone interested. He was not invited to the operations review meetings and no one read his reports. He could not help it if no one seemed to care about quality. We finished the meal in silence. The general manager was seething inside, waiting for lunch to get over and me to get lost so he could kill the Marketing person. Harry's reports were read with care from that time, he was invited to attend a lot of boring meetings, and he had to spend his lunch money in the executive dining room. There is a price for recognition. Harry and I both realized that he needed to learn more about management and communication. That was when I decided to set up the ITT Quality College.

This operating routine seemed to work in every plant. No one paid much attention to the quality manager in the "commercial" world unless they had just the right kind of personality, and none of them did. (If they had much personality or ambition, they immediately got out of quality, as it was considered a dead end.) Because of this they needed some help, both in organization and in education. In government work the contract required that the quality function report at an organizational level; that at least made people pretend to listen. Our new corporate policy would bring that about in ITT. However, it was apparent that we needed to organize the quality managers so they could become a force, and that we needed to educate them.

The financial people, from timekeepers to accountants to company comptrollers, all reported directly to New York. The local general managers had nothing to say about their activities, policies, or even who they were. The purpose of this was to make certain that no one was fooling with the numbers in order to make something look different than it was. The corporate comptroller was a senior vice president, a member of the board of directors, and went to all the important meetings. I was sharing a secretary with two other guys, had the worst office in the building, and zero staff. I could only hope that the comptroller had begun the same way. We were going to have to do quality in a much different manner than finance. Actually, I would have refused to have all the quality functions report directly to me anyway. The general managers would be calling in to say that "your dummy sent out some very bad stuff—what do you intend to do about it?" Let them be their dummy, not mine.

In order to turn the quality managers from observers into commandos, I decided to assemble them in Quality Councils. This would provide a place for education, fellowship, and communication. The Quality Councils began with the Aerospace and Defense group. I invited the quality managers of that group, about a dozen people, to come to a meeting at one of the Los Angeles–area plants. Being in government work they were used to going off to meetings and used to being on committees; they picked up on the Quality Council concept right away. I selected one of them as chairman, making it clear that we were going to do this throughout the corporation in order to install prevention as a normal way of life. The quality function people would become very important and useful if they cooperated with me and learned how to get the right things done. Lack of cooperation could produce the opposite result. All of this sounded reasonable to them, and we began to lay out plans for that group.

The "commercial" people were another situation all together. When I sent out a notice inviting the quality managers of one group to a meeting, I was deluged with calls from their bosses saying that they could not be spared from the plant. The concern was that nothing would be shipped if it was not possible to get waivers on command. I also got a dressing down from my boss who said I should be out solving problems instead of having quality meetings. My response was that I was preventing problems, that our worldwide quality operations were kindergarten grade, and that he should recognize that one guy traveling around the globe was not going to solve anything except some airline's cash shortage. I told him my vision of a stewardess coming to my seat one day when I became 65, to pick up my air travel card and hand me a gold watch. I did promise that every place I went I would solve at least one current problem and save them enough money to pay for my trip. I showed him three cases where I had obtained written agreement with the comptroller that my suggestions had saved enough to pay my present salary for the next dozen years. There were so many opportunities it was like picking up seashells, but the regular operations needed to learn how to do that on their own.

When we began to have an understanding I talked to him about the "Quality College." We needed to teach the quality professionals and start on the executives after that. We would have classes on how to manage quality and bring all the managers to school in groups of 20 or so at a time. They would learn about management in general, how to deal with executives, and how to install a quality improvement process. The goal would be to make ITT the standard for quality, worldwide. If that was going to happen, we needed someone to help me; I wanted to bring Bob Vincent up from Martin for that job. To my surprise, my boss agreed to hire Bob on the spot, but he said I would have to deal with Personnel because they were impossible. He would sign the requisition but the rest was up to me. Hiring me had taken about six months, and most of that time was used up in happenings within ITT Personnel, although some time was lost when an executive vice president got fired.

Personnel was a thorn in my side for the next 14 years. It wasn't that they were picking on me; they just had their own agenda, which I think they imported from Jupiter or some other abstract place. They had developed a complex series of tests and evaluations, along with an extensive interview with one or two psychologists. All of this information wound up in a blue bound booklet that laid out the applicant's assumed personality, prognosis, and history. This was presented to the executive doing the hiring who then used it to make a decision, or not. There was no bottom line on the report, no "yes" or "no." My observations over a period of time was that about half of the senior people hired did well and the other half quickly left for one reason or another. I always felt you could tell if a person was not going to work out in a few moments of face-to-face conversation. But to do that it was necessary to understand what it took to get along, and be useful, in an organization like this. Neither the Personnel people or the psychologists had any way of knowing this. They did not relate to the organization or its culture, they lived on their own planet. For a while I kept an unofficial scoreboard in my desk, like I did each year during the Miss America telecast. This showed that only about 25 percent of those hired lasted more than 18 months.

Asking for permission to do something is not the way to get things done in a large corporation. No one with authority will ever want to do anything different. No one will understand why you want to start something new, or worse, change what has been there for a while. Doubt will be raised about exactly whose approval is needed and how this will affect other operations. So I learned not to ask, just go ahead, keeping everyone informed. They think that you must have someone's approval and are afraid to ask who that might be. However, the things you are doing have to be presented one day and that has to be done in a manner that does not appear to threaten anyone or any existing programs. It takes a great deal of selling to convince people that your intention is to just make life easier and more productive for them, without adding any control over them. They are not used to someone wanting to be useful.

In late 1965 I went to London and became an unashamed Anglophile. My hobby has always been history, and to walk around actually touching all those treasures was marvelous. It wasn't long before I could almost qualify as a tour guide; the place is like Disney World with real people. Working with three British ITT units, I began to realize that this was an entirely different world of management. There was little contact between those with suits and those wearing work outfits. It didn't take an expert to see that productivity was very low; it took twice as many people to do about half as much. To make changes in anything, particularly if the suggestion was coming from an American, was not really possible. All the people I met thought it was a shame, but there it was. They would like everyone else to adapt.

On this same trip I went to Brussels where ITT had its European headquarters, to Paris, and to Stuttgart in Germany. My purpose was to learn what the corporation was about; my boss thought I should be getting hardware moved. He was particularly concerned about a plant in southern England that made teleprinters for the British Admiralty. They had not been able to get the machines past the government inspector and there was a big backlog. Would I go over and see what could be done?

My schedule was to head on home after two weeks in Europe. Some of the world staff stayed months on a job; it was not a good idea to start that sort of pattern. However, I changed my plans and went back across the Channel to the plant. They were glad to see me but they would have welcomed Hitler back if they thought it would get Brussels off their backs. Their problem was contained in a large room full of teleprinters, all covered with plastic sheets. The government inspector, an elderly gentleman with a great shock of white hair, came by two times a week. He looked at four machines, accepted or rejected them, and then left. Since there were several hundred of the units involved, and the plant was making many more each day, it was not difficult to see where the bottlenecks originated. They showed me, upon request, exactly what the inspection was about. It was to assure that the machines met the requirements mechanically, but that did not involve the machines actually being hooked up and working. There was a package of specifications two inches thick that related to how far apart the keys should be, and how much travel should exist between them. These had been prepared by our own Quality Department, and were not part of the contract. The contract just said that the machines should receive and transmit reliably. The measurement had to do with a certain sized message, the integrity of the transmission, and, as far as the contract was concerned, that was it.

Against everyone's advice I asked to see the government inspector who turned out to be a very nice person. We went to the pub together, had a Dutch lunch, and discussed the Battle of Hastings which had occurred nearby. When I asked him why he only inspected four units each time, he replied that those were all they showed him. What he was really interested in was whether or not the machines sent messages and got them back. He really didn't understand why the quality people were so interested in all the mechanical measurements. We agreed that he would return at the end of the week and spend the following two days with us.

Back in the plant I tossed out the mechanical acceptance requirements specification and helped them take the plastic off the machines. We set up a series of tables and put the machines, all 435 of them, in neat rows. Then we set up two input-output areas where a machine could be hooked up and tested.

When he returned, the inspector agreed to the plan, which was that he would select machines at random and we would operate them for him or he could do so himself. We agreed that 15 percent of them would make a proper acceptance sample. Over the next two days, we put 100 units through the testing system. All of them performed to the requirements. When it was all over, he released the entire lot and they were shipped off to the Admiralty.

We agreed that in the future he would concern himself with the integrity of our test rather than in witnessing a special evaluation. Anytime he felt something was lacking, we would stop and conduct special testing. That way a steady flow of machines could leave the plant. It worked that way for many years afterward. I called my boss and he agreed that I could go back home, which was fortunate since they had already driven me to Heathrow Airport.

The key to getting those machines out was to recognize that the Navy wanted them for use and ITT wanted to get them out so they could get paid. The mechanical measurements that had been created for no good reason blocked the whole thing. No one even mentioned them while we were going through the testing business, and they never came up again as far as I know. This was a case of people making trouble for themselves and their customers by getting away from the subject. The Europeans like specifications and requirements; they never seem to get enough of them. If none exist, then they will create a bunch. Once created, they never stop existing.

While studying the European operations, I concentrated on the telecommunications company that each country had. Although owned by ITT, they all were called by other initials:

SESA, CITESA, CGCT, SEL, TELCOM, and such. The telecommunications business was divided in most countries between ITT, Erickson, Seimans, and a few others. Telephone systems were all run by governments; we just made the equipment for them. But government people were in all the factories and had acceptance stamps, just like the Department of Defense folks in the United States. My goal was for them to have so much confidence in us that they handed over the acceptance stamps and just sat in their offices. At this time they were very busy and were not that happy with our quality at all. Most of the time our Quality Department people were sitting around arguing about one specification or another rather than solving and preventing problems. If you can keep the subject of quality aimed at paper instead of products or services, it lowers the level of trouble.

The quality systems inside these companies were patterned after the quality-control methods developed in Western Electric. These were aimed at detection and correction. Inside the operations there would be an inspector for each four or five workers. It was possible to see that even some modest quality engineering effort toward prevention would pay for itself quickly. The problem was how to get these companies, in a dozen nations with different cultures and languages, to decide that they wanted to do this. I had learned that the worst way to initiate change was to do it by fiat from headquarters. Subsidiary companies would find ways not to brush their teeth each day if brushing were ordered from New York.

Each of these companies had a quality director who worked at the top levels of the organization. They were mostly older than myself and had spent their lives in the business, although not always in quality. For the most part they had been engineers. Georges Borel of France, Dr. Rudolph Behne of Germany, and Dr. Enrique Blanco of Spain were the senior people. Blanco was a lifelong quality-control expert of the old school. However he, like the other two, was a philosopher and recognized that the conventional practices and beliefs had not been effective. I treated them with respect and we all became close friends. I contacted them, plus their counterparts in the other countries, and asked them to meet with me in Paris during the first quarter of 1966; they agreed. My message was that we would form an Executive Quality Council for Europe, that Borel would be the chairman, and that we would plan our work for turning around quality in Europe. I selected Borel because he was a diplomat, and French. My observation was that the Europeans didn't like each other as nationals all that much but they didn't dislike the French as much as they disliked other national cultures. I realize that comes out strange, but that is the way it was.

During our first meeting they said that their management needed a message from Tim Dunleavy, the president of ITT Europe, saying that all this quality strategy had his blessing. On the way back home I stopped by the Brussels office between planes, wrote out a sterling letter of support, gave it to Tim, and he signed it happily. As he said, he didn't have anything to lose and had a lot to gain. Tim was not one who signed such letters casually and was known to carefully distance himself from those who wanted him to join their particular crusade. He didn't get to his level of responsibility by carrying someone else's lunchbox. But he recognized that all of this was what the company had to do if it was going to survive. I asked his office to distribute the blessings document, which they did. By the time I got back to the New York office on Monday, there were telexes from all over Europe saying that the managing directors wanted to come to our meetings, and loved us. The same approach had very little effect in the United States or South America. Just being in charge of something with management's blessing got you nothing in those places. I learned to bring the Europeans over to the Western Hemisphere and let them lecture the natives. That had a positive effect.

We needed something to work with, to get the word out, to hand to people. Using my trusty typewriter, again at home, I wrote a booklet called "Quality Improvement through Defect Prevention." It covered the concepts of quality management, plus the 14 steps of an improvement process. We had to lead people away from the irrelevance of Quality Control and

Statistical Process Control which were being taught in the profession. They produced no results in terms of management action, just reams of paper, and megatons of argument.

To support this understanding, I prepared tape recordings of the Zero Defects concept in all the ITT languages. In those days such recordings were on reels of tape; cassette players had not been invented yet. The tapes were placed in little "handkerchief" boxes for presentation. It was necessary to carry a 15-pound tape player in order to have someone listen to the concept. But outside of learning a dozen languages, this seemed to be the only way. The Public Relations Department pitched in to help me get this material laid out. Rich Bennett, who was an executive vice president, became interested in my efforts and helped me get the money for this project. It didn't amount to much and, in fact, the amount made little difference; the company had plenty of money and they were willing to spend it. It was just getting someone to give it to you for something that was new. I finally learned to just go ahead and get what I needed. No one seemed to care—when we were dealing with billions, a few thousand was not much to worry about. Companies spend a lot of time and money controlling small amounts and terrorizing their people in the process. This mindset was something else I resolved to omit when I became a CEO.

Rich had made a big impact on ITT management ever since he arrived. Geneen liked to hire bright people with senior executive experience and then just sort of toss them into the pits to see what they found to do. Rich realized quickly that very few of his colleagues knew much about manufacturing so he picked up on inventory right away. The company had a great deal of inventory worldwide but did not really know how much or of what. Rich arranged for the world staff to calculate it so he could report what terrible shape it was in at the General Management Meeting (GMM). Ed Schaffer, the director of industrial engineering, on the same staff as myself, was in charge of all this. As Rich's efforts drove the inventory down, CEO Harold Geneen approved and soon Rich had other things to do, moving into the office of the president. Rich decided that I was doing the right things about quality and became a mentor and supporter. We advised each other on many things, but Rich was the master at handling situations.

He developed the business of taking a company plane and a crew of people to go visit several units. Other senior management never got to any of these places, so as a result Rich knew more about what was going on than anyone else, except possibly me. I made a point of visiting several units each month as well as meeting with quality councils. They gave me the real story on situations. Also I went with Rich on his trips.

We worked out a strategy that helped the quality effort in each place. My sources, through the quality council and my personal relationship with the quality managers, let me know what was really going on in the unit, good and bad. I would prepare Rich and in return he would ask for a comment from the quality manager when the inevitable on-site operations review began. Since that person was hardly ever invited to the review, this caused embarrassment. The lights would go back on, there would be a short break, and when things resumed there sat Rich, myself, and the quality manager in the front row. After a while the idea began to reach out ahead of us, and the general manager could always be found with his arm around the shoulder of the quality guy.

When the seven U.S. quality councils were formed I brought them to New York one group at a time to have lunch in the executive dining room with Rich, or Tim Dunleavy, who became president after leaving Europe, or even Hal. After introducing them to one another, I would leave the room so they were alone with the big boss. One side effect of this was that ITT top management would remember the quality manager when meeting the staff of a unit; in most cases they had never met the general manager before. All of this effort led people to think that they had better things to do than make trouble for the quality function.

The Management Development Department ran a two-week seminar on the ITT management system and my boss decided to send me. Later he noted that everyone else was doing something important and I was the only one he could spare. Fortunately, I don't need a lot of love from the business world. The course was run by consultants from Harbridge House and was time well spent. They actually did teach the ITT system, but in passing they played psychological games that were very helpful to me. The meeting was held in a club just outside Atlantic City. I arranged to play golf at the crack of dawn each morning, all by myself with a caddie. This let me get to the opening exercises on time. We did the same thing in Brussels during meeting week. Some of the group played tennis; a few of us went to the Royal Belgian course and swept the dew.

The first seminar day of the system program was spent with everyone arguing about the difference between "line" and "staff." This effort was particularly useless but let them get acquainted. Since I had no idea such seminars occurred at all, this one gave me a good understanding of how to teach executives in a way that would be remembered. Two weeks was much too long for participants' attention span, but I had not had any time off for so long that I really enjoyed it.

The Quality Councils suggested that we needed a movie to explain what Zero Defects was all about and I persuaded the administrator to slip it into the department budget. As soon as that was done I contacted a filming company—this was before videos—and we went to work. My experience at Bendix had been a great help with this. The producer-director and I both wrote the thing; he hired professional actors whom I recognized from commercials; and soon it was in the can. The title was *Zero Defects, That's Good Enough.* The film was very well received and we made the narration voice-over rather than dialogue so we could dub it into all the languages in which ITT had units.

The ITT management system was based on figuring out the five-year financial goals of each unit early in the first quarter of a year; then these were turned into a business plan, which was then discussed and resolved by the fourth quarter; then the staffs monitored compliance. When troubles were foreseen or appeared, the staffs swarmed in to help. It was all right to have problems; it was not all right to ignore them or not ask for help.

Hal was a terror when he discovered that information had been withheld or colored. Every month the world status was covered in the General Management Meeting (GMM) and then all the senior people went to Brussels to meet with the European executives in the European Action Committee and a bunch of smaller groups. They did this every month, and I did my best to duck out of the EAC trip. It was hard to do that stuff and still do my job. Not being an officer yet, I was not invited to the GMM; my boss went and slept through most of it.

The General Management Meeting was the heart of the corporation. It happened in New York each month for two or three days. Every general manager and staff executive in the corporation wrote a status letter according to a format. These were placed in two leather notebooks and sent to each of the 80 attendees. The meeting was held in a large conference room on top of the building at 50th and Madison. All I knew about it was that now and then there was an assignment about a "quality problem" that I was supposed to do something about. Through relationships built up by the Quality Councils, I could call the quality manager from that unit and get the straight story. They knew that I would take time to understand what was really going on, and that I would not make them look bad because of the problem. They also knew that I would find out eventually what was happening even if they did not tell me. They did not want to be put into that situation. I couldn't fire them, but I could make them wish they were somewhere else.

When executives talk about quality problems they mean that something didn't perform correctly. They do not mean that the Quality Department itself was the culprit, but other people think that is what they mean. So it is essential that corrective action requests spell out clearly how the problem was caused, how it will be fixed, and how it will be prevented. In the process, it is necessary to make certain that individuals are not targeted. "Fight the problem, not the people," said Hal Geneen, and he was correct. Every problem was actually different than it had been described at the meeting, which meant that the information received by the senior people was not complete. From this I learned not to prejudge, or get upset, or even frown when a problem came down. There was time enough for that later. Also nothing was ever as bad as it had been portrayed.

All my dealings with the senior executives were positive. They were very receptive to the actions I was taking and had complete assurance that I was somehow going to solve all their "quality problems" for them. This was not what I had in mind. But they only reacted to money, that was their primary concern, and the way executives are measured. Herb Knortz, the comptroller, responded to my request to figure out the Price of Nonconformance and had some of his folks work with me. When we were able to identify that at least 20 percent of revenue was spent doing things over, I realized that we had a tool that would focus the operating executives. I explained it all to Tim and Rich but wanted Hal to get the message. So I laid in wait for him one morning at the elevator. He always came in about 10 a.m., but he didn't go home until after midnight. When he arrived I hopped in the car with him and raised my hand to keep him from starting the conversation. He did both parts usually. I gave him what I have come to call my "elevator speech."

"If we could learn to deal with quality as conformance to requirements instead of goodness; work on prevention instead of detection; and use Zero Defects as our performance standard instead of Acceptable Quality Levels; and measure quality by the Price of Nonconformance instead of indexes—we could reduce our costs by 20% of revenue."

We arrived at the 12th floor and he got off, I went on to the 13th floor. I had just poured myself a cup of coffee when Tim called and invited me to make a presentation at the next GMM. When that took place I watched the eyes of the group executives and division presidents. They lit up when they saw that they could actually make money by learning how to do things right the first time. From then on we dealt with the real world.

In mid-1968 I was asked to speak at the European Organization for Quality Control (EOQC), who dropped the "Control" a few years later, which was holding its annual meeting in Madrid. We arranged to hold the Executive Council meeting there, inviting the American Council chairmen to participate also. We held our meetings for two days before the EOQC. When I talked about the ITT corporate program at the conference I was surprised to find my comments challenged by several U.S. quality leaders. They said that causing a company to try and get Zero Defects was wasteful and gave people an impossible task. The most vocal of the dissenters came to the platform, as was the custom there, to ask me a question. Before he could do that I asked if he would mind telling me what Zero Defects was. He described it as a worker motivation program that deceived management into believing that they could get better products just by getting the workers to sign pledges. I got up from the table and went over to him, put my arm around his shoulder, and said that if that was what ZD was, I didn't want any part of it either. To this day I have never figured out why he kept insisting that we have to do a few things wrong in order to be human. The influence of these people has been destructive in this way. Quality professionals are so busy controlling variables they have trouble getting around to the real actions of value.

My old boss had left the company and they brought in a new one. We had many people on the world staff who could have run the job, but that meant they would then have to be replaced. It was easier to go find someone to come in at the vice-president level and run the staff. That explains why many corporations lose good people while gaining ones other companies are delighted to see leave. This new boss was dimmer than the previous, in my opinion, so I decided that this was not the place for me. The quality management system was in place; the Quality College was operating; I could leave with a clear conscience. My salary was slowly edging up, and I had received a large incentive compensation each year. However, I was still making under $40,000 which was just pocket money in Greenwich. There were no stock options at my level, and I was flying tourist class across the ocean and back almost every month. My determination was that they did not appreciate me, even though they said they did. My new boss let me know that this was about the end of the road for me; he wasn't sure I would be getting a raise this year since I was pretty high already. His idea of high and mine differed. Like most executives who had been raised in manufacturing, he considered people in the quality departments to be subhuman necessary evils. He had a real concern that someone like me would be working at this exalted level in a big corporation. I thought I should be at his level or above. The clincher for me came when he suggested to me in a moment of confidence that I really was bright enough to have a good job in the company and should talk with Personnel about that. All the top guys were great to deal with, but they had no idea of what was happening down in the trenches.

The Sunday paper had a display ad in the Help Wanted section (I read everything) about this corporation which was looking for a vice president of quality. I sent them a copy of *Cutting the Cost of Quality* which had been published in a new edition the previous year. We used it in some seminars I conducted on vacation days in order to improve my income. The book didn't sell a lot and was made fun of by the quality press, all of which only convinced me that my way was the best.

The next week I received a call at home from the headhunter, Elmer Davis, who was delighted that I was interested. He had even heard of me. We arranged for me to come by his office on 42nd Street, which I did. He gave me the most detailed interview I have ever had in my life; when he was done there was nothing left to know. He passed this to his client, which turned out to be RCA, just up the street. After a few interviews at that company we were positioned for an offer. Going there would have been stepping back in time, beginning all over from a corporate culture standpoint. However, it was a much less complex company and was not growing like ITT. My plan, anyway, was to go out in the world and start a quality education and consulting company as soon as I could. I just really didn't know how to make it come off yet. I knew how I wanted to run it so that the employees, suppliers, and customers would be glad they were associated with it.

My friend George Schmidt, whom I had known from Martin, had set up his company, called Industrial Motivation, to run quality motivational programs. He had an office on Madison Avenue. We ate lunch a couple of times a month to cheer each other on. When John Delorean was running Pontiac he asked me to come and speak to his staff about quality. I went up and did that, getting them interested in Zero Defects as a management concept. This was in the days when automakers considered eight or nine defects per car as normal quality. When John was interested in doing something, I recommended that he let George, who had been on the original committee at Martin–Orlando, conduct his communication program as a consultant. Pontiac did very well for a while, dropping defects and warranty costs significantly, but ZD concepts never penetrated the mother corporation (GM) at that time. When John moved to Chevrolet, it all quit.

When RCA was about to make me an offer, Elmer asked me how I thought ITT would react. The thought had never crossed my mind. I said that they would wish me well and, I hoped, let Bob Vincent take over the controls. There was nothing else I could do for them working from this bunker where I was held away from the real thought leaders. He made me promise that if they convinced me to stay I would find him a replacement. I agreed, knowing that it was no problem. When I told my boss that I was going to leave he said I would have to go see Dunleavy, the president. I couldn't get near him, or Rich, or Hal. The Personnel people wouldn't talk to me; there was no one to quit to. It was starting to be the holiday season, there was snow and ice, and my family and I were all getting nervous about the situation. So we packed up the kids and drove to Orlando for some warmth and friendship. We checked into a motel with a heated pool and sat around thinking about what to do.

The second day there, the phone in our room rang and it was Hal who was in Brussels. He apologized for interrupting my vacation, said that he would like to pay for the family trip because of it, and informed me that he hoped I would stay with the company. He said that they would make me a vice president reporting to the Office of the Chairman, raise my salary to $50,000, give me a large stock option, and put me on the executive bonus role beginning immediately (which meant the whole year). He said that he had been receiving calls from the unit executives telling him not to let me go because I was the only effective seagull in the place. I was overwhelmed and mumbled "yes"; he thanked me and hung up.

Rich told me later that Hal had said I should have told them I was unhappy, but Rich replied that it was their job to know that. He knew I would never speak about it. However, it is true that one of the main jobs of the senior people is to know the attitude and feelings of those down the line. I was not quitting because of money or even position; it was just because I felt they didn't need me. Sounds kind of silly for someone who is supposed to be tough about it all.

I had just finished writing a book called *The Strategy of Situation Management*. The publisher was about to put it out with an RCA résumé inside the cover. I had to stop that and did, but about 2000 had been distributed. Every now and then I run into someone who has one of those with the wrong résumé in it; they ask me to sign and I pretend I am going to swipe it. It is a collector's item, although hardly like the stamps with the upside-down airplane.

After everything quieted down, one of my colleagues went to the Office of the Chairman and announced that if he did not become a vice president, "like Phil did," he was going to quit. He was sent on his way home that day. This, and a couple of other occurrences, made me realize that what I had done produced a good result only because it had been done without the intention of causing that. The ITT folks did not feel that I was blackmailing them. I found out later that Hal had called Mr. Sarnoff about my situation to make sure that the two corporations were not going to get into a squabble.

I was elected an officer of ITT at the December 1968 board

meeting and invited to the annual Christmas luncheon. Many of the units had sent little presents with their logo on them to be distributed to the officers and directors. These were identification items like key chains, flashlights, clocks, and such, nothing valuable in themselves. They keep popping up in drawers and boxes to this day. The baking company sent each of us a fruitcake. I am still searching for someone who actually ate some of a fruitcake. Over the years I have developed the theory that there are actually only a dozen fruitcakes in existence, they just keep being sent around. We used ours as a doorstop.

Tim took me aside at the buffet, and asked for a proposal of the staff and budget I was going to need in order to do the quality job properly. The result was a request for four people, two secretaries, and a Quality College coordinator. It was quickly approved; we were the smallest staff in headquarters. But we became the best known.

Hal liked the ZD film and wanted the board to see it; he felt that it was time to begin taking quality seriously. So I was scheduled to attend the next board of director's luncheon to make a short talk on quality, and show the film. Tim suggested that I write out what I planned to say so they could look at it before the session. This was not my usual practice. I never made a talk without thinking about it and actually writing it down. However, I never wanted to be held to it, and certainly would never read a speech to an audience. I much prefer to calibrate the group before getting into the text. What is right for one may not fly with another. But in my tradition of giving unto Caesar, I wrote out a couple of pages and sent it to all three of them. Tim and Rich both gave suggestions, which I ignored, but Hal said nothing. Communicating with him was like talking to Mars. Your message went out and when you thought it had been lost in space a reply, or even some action, would occur. He had his own time warp.

At the board luncheon I had an opportunity to listen to a group of heavy hitters chat about the world situation and about the inside of business. They all knew everyone and it was a delight to just sit there and bathe in this new environment. After lunch Hal stood up and made a few comments about me and the effect I was having on the corporation. Then, with no notes or anything else, he proceeded to give my speech. It had registered on his memory and he just rattled it off with, I'm certain, no idea of where it had originated. When he got to the end in about five minutes, he signaled me to take over, but my mind was blank. I rose, smiled at everyone, and signaled the projector operator to turn on the film. By the time it was over I had written a new speech in my head and it all turned out right. After that he invited another officer to make a presentation at the next meeting, but it did not work out well because the person rambled on and made a poor impression. That is something no one wants a board to go away remembering. So John Monaghan was hired to prepare executives for these presentations and from that date on, someone was there each month.

It is a cliché to say that business executives are usually not good speakers. It is also true. When an important person speaks to those over whom they have power, it does not make much difference how proficient they are about it. But when that same person speaks to strangers or other nonsubordinates most of the impression comes from the way they appear on a platform. Unfortunately, very few realize this and no one wants to tell them. My observation has been that they can improve with coaching, such as John provides, although they never really get very good at it. They become adequate and avoid embarrassment, but success eludes them because they never really thought they had a problem in the first place.

ITT—APPLYING THE CONCEPTS

By now my ITT life was becoming somewhat routine. The Quality College ran continually somewhere in the world, improvement programs were happening in most companies, and top management took their responsibility for the company's integrity seriously. The early battles about things being "good enough" in the General Management Meetings had disappeared. Instead of arguing about quality, or blaming the quality function for problems, the executives understood their personal role and insisted on a standard of Zero Defects. They also realized that this all made their jobs much easier. They had time to concentrate on the future rather than having to spend time deciding if the past was usable.

The Ring of Quality program was working well worldwide. People were asked to nominate someone, other than their boss, as the person they recognized as their personal example of quality performance. Each year the president would present about 25 rings in dinners held in the United States and Europe. Special presentations would be held in other parts of the world. Senior executives who really got wrapped up in the quality flag would receive special Ring of Quality awards from the Executive Councils. The effect of this award program was positive in every way. No one complained about those who received rings because they had been peer nominated. Those who were nominated but did not receive rings were given special lapel pins and a certificate signed by me. This was usually presented by their general manager at a luncheon.

We made a new movie, *Why Me?*, which showed an executive trying to blame everyone but himself for his problems. I took care to make certain that the subject and dialogue were understandable to the Asians and Europeans. We used recognizable TV actors and it was very well received. There were no VCRs in those days, so everything had to be done by way of 16-millimeter projection. Bob Weimer, the producer-director, and I fought tooth and nail over the concept but cooperated smoothly on the script. The film nails executives directly with no room for rationalization.

Systemwide, things were much better in the matter of quality now; everyone was aware of it and took their personal responsibility for action. The quality-control professionals who could not adjust to prevention had been weeded out with a new, less technically oriented type of quality management coming in. The Executive Councils were a pleasure to be with; in Europe we had wonderful meetings complete with great meals and fellowship. This was a large move forward from the early days of suspicion and confusion. At one of the first meetings, I was told that I would have to work very hard to understand all the concepts and concerns that the various quality directors had. My reply was that to the contrary they were going to have to learn to understand me if they ever wanted to have a concrete policy and operating plan. They responded positively to this and we had a completely apolitical relationship for the rest of our days. They took the business of giving the customers what they had ordered to heart.

The operating people were becoming more responsible in this concern also. We didn't have people trying to slip substandard things out in order to make a date; it just became the thing not to do. The role model of the resourceful manager who could deliver regardless of the situation began to vanish. Senior management started to appreciate those who did the job properly without a lot of fuss and bother. They did not want to be snatching victory from the jaws of defeat all the time; they wanted a smooth journey. It was less painful and made a lot more money. It was also easier to manage.

Sheraton, under Bud James, had launched the first activity in the service area, and soon after that Hartford Insurance began the process. Hartford's merger agreement stated that they were to be free from ITT headquarters' involvement for a few years, but they asked for us Quality Department folks to come visit them. I always thought they wanted to show that they were not against corporate involvement but also didn't want to get involved with the hard-nosed staffs from finance and operations. Our meeting started what became a very worthwhile effort inside the Hartford.

The chairman, Harry Williams, took me to his office one day and showed me a life-sized painting of an early chairman, complete with watch fob, flowing coat, and a stern demeanor. Harry said that he never signed anything without looking up to see if there was approval on the chairman's face. He thought the old gentleman liked the idea of trying to get things done correctly the first time. These nonmanufacturing companies did not immediately recognize that they spent about half of their effort doing things over. What the hotels looked at as gracious attention to the guest was actually rework. Insurance companies rarely seemed to get the policy written right the first time. Now they have computers to help them do better, but it is still a struggle that they often consider inevitable. In those industries the price of nonconformance is 40 percent of the operating costs. The education of the executives who run those kinds of companies has some blank spots in it when it comes to understanding what is involved in actual work. They seem to spend more time trying to create a foolproof system than in helping people learn how to do what is involved with the current system.

I found the inspiration for personal creativity all about me. That was one of the joys of such a large and diverse organization. For instance, at the Mannheim, Germany, SEL plant I reviewed their cost elimination program (I had learned to teach them to do "elimination," not "reduction") and found that they had cut out $250,000 in the last year. The manager told me they had 1000 people. This computed in my mind to a savings of $250 per person, which meant that each of them had saved one dollar a day for a year, essentially. On the way back to Brussels I wrote the BAD program on a cocktail napkin which had been under my Perrier. BAD stood for "Buck a Day." If we could encourage people to look at their job and take a dollar a day out of it, then we could have a large cost elimination with little effort. My thought was that the form for suggestions should be one page only, and that the author would receive a coffee cup with the logo "I had a BAD Idea" on it. That would be the whole prize. Then someone could draw one suggestion slip a month out of a barrel and give the contributor the number-one parking spot. We would have signs that said such things as: "BAD makes cents" or "the BAD guys ride again." Real camp. A couple of the groups took up the idea immediately and ran it for the suggested 30 days. In every case there was 100 percent participation (as opposed to 5 or 6 percent in suggestion programs), and the cost elimination goal was exceeded. Some operations would not even consider doing it even though there was a great deal of testimony and proof that it was worthwhile. After a couple of years, George Schmidt asked if he could include BAD as a product of his company, Industrial Motivation Inc. I agreed, helped him develop some material, and with ITT's agreement he paid me a royalty. George still does the program almost 20 years later.

I had a heart attack early in 1972 and learned that I was not immortal after all. This shocked everyone, including me, but it did convince me to stop smoking and to begin to take wellness seriously. I also decided that I was not going to do anything I didn't want to do from now on. I would not go to meetings if they weren't going to be useful; I would not get on committees; or I would not do things that took me away from my family. While in the Greenwich hospital I didn't even call the office, I just read and watched the news. A book I had recently written, *The Art of Getting Your Own Sweet Way*, was being well received and I had a good time reading reviews that were arriving. It was a delightful experience. The whole story of the heart attack and recovery became part of my material and was included in *Quality Without Tears* when I wrote that 14 years later. In my recovery I learned that people and companies both have lifestyles that very much determine their health. They also both have attitudes that predict their success or lack of it.

I was becoming concerned about the future. ITT had an excellent pension program and with any luck I could take advantage of it in 1981 when I would be 55 years old. At that time I could get out on my own and see if there was an opportunity in quality management education. I could tell I was beginning to get bored with the challenge that remained. Still a lot of interesting things were happening.

In one of the big telecommunication plants in Europe they learned that fingerprints were showing up through the finish on the switching systems after the switches had been in use for a year or so. This was particularly true in tropical or other high humidity environments. These installations were large racks of switching units mounted in steel and aluminum frames. Although the metal had been treated and finished, this discoloration and eventual deterioration was present. The plating procedure was modified and improved, but the plant management felt that the problem was not solvable because only some people's fingerprints caused this reaction. It was thought that menstruating women were particularly prone to producing an acid on their fingertips. All that was necessary was for the metal frames to be touched; they could not be assembled if they could not be touched.

I suggested to the operations manager that they needed a "Zero Fingerprints" program, and he laughed along with the rest of the group. However, about two weeks later he began to have meetings with the employees and staff on the subject of eliminating this problem. The result was the purchase of boxes of cotton gloves, and an awareness instruction on the necessity for keeping metal surfaces clean. Visitors were handed gloves and instructed not to touch anything anyway. The operations manager became known as Mr. Zero Fingerprints, and soon the other plants were coming in to see how all this was done. The problem went away because of this effort and soon it became a normal part of the way things were done. It was sort of like wearing safety glasses.

The hardest part was getting top management to cooperate. They never touched anything in the plant anyway and were quite stubborn about wearing gloves while visiting the work areas. But the operations manager invited a group of the New York executives over for a visit during the business plan meetings, and they all got off the bus wearing gloves. That wiped out resistance.

In the mid-1970s, because of some unfortunate political events, ITT's stock price was lower than it should be, and there were some significant problems rising. Hal seemed, to me anyway, to have lost a little zest for the complexity of the organization. He began to delegate more doing but never took his finger out of the pot. A new electronic switching system was burning up a lot of money in the development phase and was never going to get off the ground unless stronger leadership was applied in that area. But it is very difficult for a young tiger to emerge in a structured organization. What I was doing was successful only because it did not threaten anyone or step on any ground with footprints already planted. The troops were getting restless; several of the brighter officers had gone to be CEOs in other companies. Many were looking around.

Most of the officers and other higher-level people owned a special stock deal where the money was borrowed to buy stock and the dividends paid the interest. Unfortunately, each of them were at too low a price, even though they had been issued at 50 percent of the stock price at that time. They were counterproductive as were the regular options on which the government was changing rules concerning cashing in on anything. But everyone kept working hard anyway—the ITT people were the hardest workers I ever saw. I worked hard also, but kept regular hours, stating firmly that my family came before the company. This was thought to be an achievable lifestyle only because I was considered to be eccentric anyway. Once I was questioned on it by one of my bosses who said that it didn't present a good image, the business of leaving work on time. I asked if there were any departments anywhere more effective than mine, and he had to admit that the answer was no. After that we dropped talking about it, but people did notice. You can get to thinking that you are doing a lot if you put in long hours. However, much of the time executives in all companies concentrate on things that make little difference. Exhaustion is not always an indication of results.

In 1975 I joined a trip to the Far East with ITT President Tim Dunleavy and Jim Purdy who ran ITT out there. We took the 727 and a dozen people. It was an interesting trip for me; I had not been out there since the end of WWII.

In Japan we dealt with a bakery which was doing a joint venture with Continental Baking, and with the business community. At a reception I found that I was a celebrity in Japan. Men came with translations of *Sweet Way* and with articles I had written on quality management and Zero Defects. I signed autographs and even tried to do it in the Japanese script on the front of the books. They all asked if this was my first visit to Japan. Since my previous one had been as part of the occupation forces, I usually lied about it and said yes.

I had the opportunity to slip off and see a couple of factories. The main impression I received was that the general managers explained their quality process and plans to me rather than delegating that to the quality manager as in the United States and Europe. That spoke to me of the future. All of the companies had quality policies and practices and were very involved in training and communication programs. They thought that my ideas on quality management were very advanced, and knew that I was not very appreciated by my colleagues in the United States. They all were working to the goal of Zero Defects. They had a hard time understanding why anyone would want to start out with the idea of making things wrong. They were aware that most U.S. and European executives were woefully ignorant of their responsibilities about quality. They were delighted about this and hoped that no one would ever wise them up.

We went to the Philippines and saw Ferdinand Marcos who had more security than I had ever witnessed before; then on to Taipei where I got to speak English to 300 children in a museum, one at a time. When I said "good morning" to a little girl and she responded in English, all the kids immediately lined up behind her. Then each of them stepped in front of me and said "good morning, sir," we shook hands, and I responded "good morning, young man (or lady)." It took over an hour but I got a big kick out of it. Taiwan is a poor land supporting a lot of industry but the people don't get much out of that, at that time anyway. The hotel room was bugged, which made the ladies uncomfortable, and it didn't thrill me either. The people who planted the bugs weren't even subtle about it; there was a foot-long microphone sticking out of the ceiling. I often wondered what they did with any information they gained.

Hong Kong was worth the whole trip; we shopped until there was some question as to whether the plane could get airborne afterward and played golf on a course where the fairways crisscross each other. In Australia we found a lot of ITT people and had a conference, more Ring of Quality presentations, and learned a lot about why it was so hard for the company to make any money there. The market was all the Asian countries, but the Aussies did not like Asians much and so did not build up relationships. Australians are charming and interesting people, but it sure is a long way down there. They say that no one comes to Australia on their way to anywhere else. On the way back we stopped at Fiji for fuel and spent enough time to realize that the main occupation there was going to the airport and watching the jets come and go.

I realized from this two weeks of exposure that I was going to have to include these areas in my thought process if I ever started my own education firm. There were international companies everywhere; we would have to deal with them as they lived in these cultures. I began to read the newspapers and magazines from that part of the world and to plot my strategy for the consulting business. It began to look more like my business would be education rather than standing around offering advice. Everyone needed to learn.

The Spanish, Italian, German, and Belgian companies were doing well with quality improvement, but things were moving slowly in France. The French companies decided to get me off their back and agreed to try out Zero Defects, so they selected the switching plant in Laval, which is about 90 miles from Paris. I always thought they picked Laval because it was close enough to show off if the concept worked, and far enough away so no one would ever know if it failed. The plant manager, René Peyreguer, came to the United States and lived with me for two weeks. We went all around to the ITT operations so he could see how the quality improvement process worked. René had learned to speak English from the Germans as a prisoner of war. So when you said something to him he translated it into German and then into French. Reversing the process brought an answer in a little while. Once when we flew into Dayton for a meeting, I told him that "here lived the two men who invented the airplane." His reply was that "in every city in the world lived the two men who invented the airplane."

After the trip he went back to France and started work. His approach was just right for his area; he really understood what this philosophy meant and did not try to just paint a layer of coating over an existing lifestyle. He went about changing the way people looked at things. He realized that it was a leadership problem, not a matter of applying a planned set of techniques and actions.

René met with all of the employees in groups of 25. These were rural people, primarily young girls, and they were used to the conventional way of working, which almost always required a certain amount of reworking. Their prime component was a stacked relay which then was put together with its brothers and sisters in the larger frames and the whole thing was wired together, one terminal at a time. The rejection rates were very large and rework was well organized. The young women workers were in a culture that had them taking most of their wages home to their parents or husbands. If something went wrong, the parents were the ones to be called.

René started the only incentive program I have ever seen work. The plant would pay, off-line, a small amount for each relay that was found by the inspectors to be completely correct. Those which were not defect-free would be returned to the worker for correction, and that person would receive a minimum rate for the time spent doing things over. It was to the workers' advantage to get with the movement. Most of these girls were required to turn over their paychecks to their parents each week. This off-line money was theirs to keep. The profits of René's operations doubled each year for the next several years. However, this did not lead the other French plants to get much interested. They did the system things and benefited from it, but management just would not get that close to the people.

As the 1970s passed by, I was becoming concerned with the job and its future. Several opportunities had come and gone within the company because it was felt that I was unique in this position. In reality, I did not want to become executive vice president or even president of the corporation. I thought that these jobs were mostly reacting to the problems other people caused and had little to do with real leadership. Corporations could be led but they needed a culture that permitted it to happen. Also, I began to realize that I didn't really make much money compared to what some of the group executives were receiving. I began to feel unappreciated again, which is the proper process if one is to determine to make a change.

The only thing I could do was to set up my own firm. But I had no desire to go building to building asking companies to be my clients. The problem was to figure a way to have them come to me. The answer had to be a book that executives would read. I asked McGraw-Hill if they would be interested in publishing it, and they gave me support on the idea. They had published *The Art of Getting Your Own Sweet Way* in 1972 for me, and it had worked out well. I learned to handle the TV and radio book-promotion circuit. The book sold well, and is still in print at this writing. So for the next two years I lugged my World War II foreign correspondent typewriter everywhere I went. All of the writing I had ever done up to that time had been away from the office during off-hours, because it is just impossible to write there. I can't remember ever writing anything for publication during office hours except company memos and essays. Since I did not drink anymore, I had plenty of time on the road to sit in my room and write. I always have said that nondrinkers don't have anyone to talk to after eight o'clock while traveling. Very true. I also wrote on weekends and in the evenings if the family was doing something else.

In the years between writing my quality management books, I had been an actual executive, living with others, and answering for any crimes that were committed. This gave me a view of quality that did not exist in any other writings. Those who wrote and taught about quality management, quality control, quality assurance, reliability, and all the other titled functions usually had no high-level operating experience. Most of them were college professors and statisticians who had advised rather than produced. They also knew very little about business management and did not pretend to do so. This is perfectly all right when advising those who are professionals, but to ordinary people these books were not readable.

American executives were coming under the impression that the problem of quality in the country was the American worker. In reality it was the management who had become separated from both worker and customer through years of success. Companies like IBM, Xerox, General Motors, RCA, General Electric, and many others really thought that they were the industry, that whatever they wanted to do, or set as a standard, would be what was. Textile manufacturers, faced with defect-free carpet from overseas, still insisted that 15 defects per hundred yards was the proper standard. They pushed for government regulation in order to stop others from being better at quality. The consumer electronics business was flowing to Japan, quickly; TV manufacturers refused to use solid-state technology or preventative manufacturing processes; automobile companies would not recognize that their 10 defects per car average was not what the customers wanted any more. Having seen that Japanese cars worked as advertised, customers were demanding the same from the American companies.

The quality spokespeople were saying that the need was to work harder at quality control. But that was not getting anyone anywhere; it just made the products more expensive. I thought I could do for the country what I had done for ITT. The company was far from perfect but their cost of quality had fallen from 20 percent of sales to well below 7 percent. Their integrity was in the right place. But my real hope was to deal with the world problem. If executives could learn to get things right, and to quit wasting resources doing things over, then there would be work and jobs for all. Anyway, that is what I thought.

So I wrote *Quality Is Free: The Art of Making Quality Certain*. McGraw-Hill hated the title, and looked fruitlessly all through the material for some charts or quality-control stuff. This was not the sort of book they had experienced in the past but they were very brave about the new one. They finally gave in and published it as I wrote it. My plan was that if it was a success, I would leave ITT and go start Philip Crosby Associates, Inc., in Winter Park, Florida.

As I turned the book in, looking forward to another eight months or so before it hit the streets, I received a call from an old friend who was on the board of the American Society for Quality Control. The president-elect had died unexpectedly and they were wondering if I would consider taking over the job. My term would be in 1979 but I would go on the board immediately, which would give me time to learn how it all operated. I asked, "Why me?" and he said that the board was concerned that the society was falling behind and they thought I might be able to lead them out of the shadows. I wasn't sure I believed that, but he reassured me that the board would give me full support. There didn't seem to be any real reason for avoiding this responsibility, so I agreed to do it. His comment was: "Gee, you don't even have to ask anyone."

This may rank with one of my bigger mistakes. I found no interest among the majority of board members in getting the country straight on quality. They just wanted to make certain that they got to do important things in the society. After assuming the job as president I brought the new team together, at ITT's Bolton, Massachusetts, country club, for strategy sessions; began a public relations activity; and suggested that we change the name to get rid of the "control" stigma, like the Europeans had done, and gain from the emphasis on management. When my book came out the Ethics Committee chairman warned me publicly that he was concerned I had taken the job just to sell books. I tried to resign but the chairman and new president-elect said they would resign also, which made the whole thing look like a Mack Sennett production. The old guard just would not look at what was happening all around them, so I backed off and waited for my term to expire. I announced that I would not serve as chairman which would normally happen, and asked to be permitted to resign from the ASQC altogether. That was not possible because ex-presidents do not pay dues. Go figure.

Quality Is Free was reviewed in a *Business Week* issue in March 1979 as part of an "America is going down the chute" article. The writer used the logic of my book as the key part of the cover story, and the review itself was better than I could have written. They said that the key to the book was the five "absolutes" of quality management. Quality had always been assumed to be goodness; I defined it as conformance to requirements. That meant doing what we said we would do; it meant that management had to take requirements seriously. The other absolutes referred to the system of quality being prevention; the performance standard being Zero Defects; and the measurement being money. Management could understand these concepts. The last absolute was for quality professionals, and I dropped it from management material; it said that there was no "economics of quality." This meant that it was always cheaper to do things right the first time. Quality Control did not agree with that, but now that has changed.

After deciding to implement my plan, I went down to the head office and tried to quit, but it wasn't until I convinced the new CEO, Lyman Hamilton, that I was serious that the Personnel Department would cooperate. Rich asked me why I wouldn't stay if they made things even better for me, and I finally admitted that the Personnel and Administration Departments had driven me out of the company. They were arrogant and impossible, like a Russian bureau. But even if they had been at my beck and call, it was time for me to go. I wanted to try my own wings.

I had some vague plan about going to see IBM or G.E., but that would just be more of the same and I wasn't sure they even wanted me. What I learned in those 14 years of ITT has filled several books and has provided me with a career that would never have even been in my dreams. I had no information on such things, so it was impossible to wish them. In 1957 I would have sold my lifetime of work for a guaranteed $10,000 a year. Hal, Rich, and Tim recognized that I could contribute to making their dreams come true, so they gave me an opportunity to create a life for myself. I have not met anyone since with their vision when it came to people. Most executives are so self-centered that they do little to grow their people. After I left the company the corporate quality job received little attention, but that worked out all right because the culture had been built into the units. They kept right on working at quality management, anyway.

The first official day of work for Philip Crosby Associates, Inc., was the first of July 1979.

While the company consisted only of myself, and had no resources or facilities, I decided that it would be run in a manner that would make it a wonderful place to work, and at the same time it would be more profitable than any other organization in the field. I made this list:

- We will take the creation and accomplishment of requirements seriously.

- All employees will be called associates, will be treated with respect, will be selected carefully, and will be given every opportunity for professional growth.

- We will teach those who instruct in the Quality College, or consult with our clients, to understand the Absolutes of Quality Management. We will hire none with quality-control experience.

- This will be a management education company, not a consulting organization.

- There will be no Personnel Department.
- We will have a pension program, thrift plan, and a fair hospitalization supplier.
- Management will be required to keep in touch with the reality of daily life.
- We will have a monthly family council meeting where everything will be open for discussion.

PCA—CREATING A ROLE MODEL COMPANY

When *Quality Is Free* suddenly became a best-selling book, I began to receive more speaking invitations and media opportunities. The executives for whom I wrote the book actually related to it and were reading it. Authors can tell when someone has actually ingested some of the words; they can also tell when the reader has not gotten the message. Quality-control professionals, particularly those who were prominent consultants and teachers, dismissed it quickly. One said: "The clown has written a book." Most people could not get past the title since they considered quality to be goodness and everyone knows that goodness is not free. Actually since wastrels traditionally spend their money on badness, one would think goodness didn't cost that much. Also, they could not relate to the lack of statistical charts and my unremitting emphasis on management's responsibility for quality. Some dismissed my work as being "worker motivation" or "exhorting the workers." I never understood this attitude because there is nothing about that subject in this book, or any of my other ones. Some of the well-known seminar givers used a good part of their time denouncing my ideas. None of them asked me any questions.

Many speech requests came just because the current popular business book is always treated that way; those who must arrange for speakers like to pick one that can't lose, at least for a while. But most of the calls originated because quality was becoming a high-priority concern for management. They were tired of losing market share, and thus revenues, because of it. ITT was very good about referring requests to me in Florida. There also were calls from corporations wanting me to help them get right with quality. "Whatever that meant," they would say. Soon I had scheduled 30 or so speaking engagements where I would talk with a company's management team. Others came in from associations and professional quality groups. The advice I received from everyone was to not charge for these sessions with the thought that it would let me build relationships. However, I have always felt that executives do not appreciate what they get for free. So I found out what Peter Drucker charged, around $2000 at that time, and I did the same. Also since my company had no money, I asked them to give me the check right after the speech. Most agreed to do this. I would take it home and hand it to my son, the accountant.

The typical meeting was held off-site, and often the attendees were in casual dress. Many executive groups plan these "strategy" sessions so they can go to a resort and get in a little golf. Personally I think that is a wonderful way to spend time. I did learn quickly that the speaker is expected to be wearing business attire and should not plan to become intimate with the people. They often invited me to play golf with them and participate in their meals and entertainment. This was not a good idea. There are so many of them that they bury you in questions and comments; it becomes exhausting. I did learn to do enough of it to let them understand that there was a warm person behind all that intensity. They could also pick up that I knew as much about the broad content of managing as they did. They did not need a translator.

Because of my ITT experience, I was comfortable in places like Pinehurst, the Breakers, and Greenbrier. The audiences were respectful and inquisitive, but not ready to jump into a completely new lifestyle. Marketing people, in particular, were suspicious of anyone who thought things could be what they were supposed to be. There was a general understanding, deeply imbedded, that quality was varying degrees of goodness and that the more good you got, the more it cost. It was also apparent that these people knew a lot about lightbulbs, washing machines, copiers, pumps, and other hardware products but had little idea of how people operated. They were fixed into strategy planning, financial management, return on investment, and the other tools of conventional industrial management. They often said that if they could run their factories with robots, that would be their preference. All of this resulted from their corporate cultures and the attitude of the big bosses.

This attitude, which turned out to be very destructive, is what I call "mainframe thinking." When companies think that they are the industry, that the customer will follow their lead without question, and that they only need to do more of the same, cheaper or with more technology, then they are headed downhill. IBM visualized workers hardwired into a central computer system; Xerox wanted everyone to come to the third floor to get copies; Sears figured that people could get all their wants from a 10-pound catalog; General Motors wanted to make just a few kinds of shells and platforms and rearrange tail fins; G.E. was making 50 engineering changes a month in a product that was 20 years old. This attitude was carried out of the big companies to infect smaller companies as managers changed jobs.

It became possible for me to know early if a specific company could change its thinking and its culture in relationship to quality. This objective was usually accomplished by a private conversation with the chief. Those who asked questions had a chance; those who gave me an hour lecture on quality were going to be dead in the water. I often feel that something happens to people's hearing equipment when they hit the top levels of an organization. They begin to feel that they must generate all the new thoughts and are busy doing that while someone is trying to present them with a new thought.

If it was obviously impossible for me to help, I would tell the CEO that we could get together again when they were ready to deal with the real world. They could be so stubborn about thinking that all was well, that even showing them their company costs of quality did not make an impression. Usually they just wanted a better class of customer. During my speeches I would kid them about this attitude and the "brain damage" they had suffered. This approach moved many of them to investigate my philosophy further.

There were two types of companies in those days: the arrogant and the searching. When I look at the list of corporations who are in trouble today, they are the ones who were arrogant at that time and would not change. Without exception, they thought that they alone really understood the customer and the market. They thought that the workers' attention to quality had deteriorated and that everything would come back if only a "level playing field" could be established as far as competition went. During my talks I would explain clearly that this attitude was impractical and that those who would not change had no future. Several senior executive teams made trips to Winter Park for a private one-day understanding of this concept. Most of them went back convinced that they needed help, sent their management to school, and got on the right track. However, those who had a big training department or quality function usually decided that they could do this themselves. It all sounded so easy. As a result they wasted many years trying to change the people rather than managing differently. When these folks write books about their experience, and relate that it takes a long time, they do not state that they were the reason it took so long. They worked on the wrong stuff for several years because they were too proud to let someone show them. The companies where the chief executive recognized that a quality improvement process was not a matter of techniques or clichés, and took personal command, were successful. The ones where it was delegated to the Human Resources or Quality departments just puttered along. There has been a lot of activity but few solid results and few permanent culture changes.

When I went to visit client companies I always made a big deal about how the top management had become isolated from their people and from reality in many cases. I grumbled about executive dining rooms, insisted on meeting the union officers, berated the quality-control people for having low standards, complained about housekeeping, and generally tried to stir up the place. It usually worked. My experience in business had been so broad that there was very little that I had not been through before. This meant that it was very difficult to fool me or to show me something that did not have a soft spot in it. Even those companies that were doing something about quality were usually just "the best of a bad lot." As a result of these visits, many executives opened their minds and caused a great deal of progress. Many just became more firmly entrenched in their way of doing things.

One of the most amazing people I ever met called one day, when the College was about a year and a half old, to say he had read *Quality Is Free* and wanted to start his company down that road. This was Roger Milliken, the CEO and owner of Milliken Company, the textile firm. Roger had to be the hardest-working executive I ever met. He arranged for all of his key people, 241 of them, to meet with us at Pine Isle, Georgia, in February 1981. We had a two-day intense session on quality management. Everything in that company was planned and scheduled right down to the last item. The meals happened on schedule, and everyone was there to eat them. The sessions began and ended on time; everyone was attentive. Even though we were miles away from civilization, each of the attendees dressed in a business suit. Milliken Company was known as having the best quality in the industry, but their management recognized that this was not enough. They had to forget the industry ideas about quality levels and learn how to improve yields by eliminating and then preventing error. It was a joy to see a group that didn't want to argue about it; they just wanted quality to be free.

They all came to the Quality College for a week each over the next year. President Tom Malone drove the thought through the company, and as a result they are the leaders in their field. He attended the classes with many groups and worked with them after hours to get the message through. They have shared their approach with hundreds of other companies and are always careful to let them know where it all began. We always called Roger our "salesman of the year" because he sent us so many clients.

As I discuss the development and growth of PCA in the context of my life, ideas, and learning, it really is not appropriate for me to discuss much of what went on with clients. There are anecdotes that serve to help our understanding and I will relate them; also I will not talk much about the individuals of PCA by name and their activities. The company is still operating successfully and many of the original clients are still involved. I have not checked the files, purposely, so this list is from memory. However, our company newspaper, which was begun when we had about 30 employees, is a good memory jogger when it comes to events.

Concerning clients, I have always been very proud of the fact that we did no sales work, never made a call—they all came on their own: IBM; Xerox; Tennent; Corning; Milliken; Motorola; Mostek; Cluet-Peabody; J. P. Stevens; Cellulose; General Motors; 3M; Brown and Root; Chrysler; Johnson & Johnson; Federal Prison Industries; Federal Reserve Board; Clark Equipment; AT&T; Armstrong World Industries; ICI; Seagrams; All-State; Navy Yards; Westinghouse; General Electric; Armco; British Petroleum; ARCO; Amdahl; Owens Corning; Copperweld; Savin; TRW; Hinkle; Herman Miller; and several hundred others. In some of the larger companies, and all of the smaller ones, it was the CEO who drove the effort; in others a single division or group was the participant. Whatever the source, six thousand executives and managers a year were attending the Quality College before long and many thousands more people were utilizing the tapes and other internal material we developed and taught.

As I look at some of the books being written today, I recognize a great deal of my material that has been taken without saying from whence it came. But most people who talk about serious quality management, as opposed to TQM and other shallow efforts, make proper attribution. A couple of our client companies took the material they got from attending the Quality College and started up in competition with us; one took the people we sent to consult with them and helped them set up a consulting firm. After a short while they fell out and sued each other over the material I created at that. I could have sued everyone, and the lawyers encouraged us to, but I felt they deserved each other. However, none of these were ever serious competition; they concentrated on price rather than results. Only one company stiffed me on the speech fee; they said they thought they had given me a great opportunity by having all their people come to a meeting. Those few companies who did what I considered unethical things came back to us for help later on and were immediately refused. How's that for a forgiving attitude?

As I started to put PCA together as a company, I wanted to establish a strategy that could be understood. If this was to be a real company one day, then it should be designed so that people would be proud to work there and clients would receive the most help they could stand. For this reason the name Philip Crosby Associates was chosen deliberately. Most organizations like this would have been Philip Crosby and Associates. It was very important to me that the employees feel they were an important part of the operation. This seemed to be a way of saying that. To my knowledge no one else ever thought much about the difference, but it was always a reminder to me.

During this strategy phase I ran across a comment by Warren Buffett describing a "wonderful business." He listed the following criteria and I compared PCA to it:

1. Good return on capital (PCA would require very little capital.)

2. Understandable (We would educate management and help them educate their people. That is understandable.)

3. They see their profits in cash (We asked people to pay before they came to school—no credit, no discounts.)

4. A strong franchise and thus freedom to price (No one ever complained about our prices, except the quality-control people, and we did not deal with them.)

5. Does not take a genius to run (Within two years everything was delegated by me except the creative director's role.)

6. Earnings are predictable (This was true, but revenues turned out to be subject to recessions, where companies cut back on learning, and not predictable.)

7. Not a natural target of regulation (No one cared about us one way or another. We paid our taxes and did not want the government as a client.)

8. Low inventory and high turnover (We had no inventory except a few tapes and classroom materials.)

9. Management is the owner-operator (All employees had stock.)

10. The best business is a royalty on the growth of others (We set up clients to have their own colleges and pay us a fee per student.)

Taking all this with what I had learned over the past few years, I wrote out a strategy at the end of 1979. I had spent enough time with clients, at the top level of the companies, to know that the executives were ready for change and that we were the ones to cause that. The management had to understand that they were the cause of the problem, and the employees of the company all had to understand quality the same way. What was important were the concepts, not the techniques. Statistical Process Control, Quality Circles, and other popular programs had nothing to do with the cause and effect of quality. They were just tools, and properly applied, could be useful.

I built the Quality College around the Four Absolutes of Quality Management as they had evolved for me over the years:

1. Quality means conformance to requirements, not goodness.

2. Quality comes from prevention, not detection.

3. Quality performance standard is Zero Defects, not Acceptable Quality Levels.

4. Quality is measured by the Price of Nonconformance, not by indexes.

Dean Martin Schatz of the Rollins College business school (Crummer), in Winter Park, offered us a classroom in return for a donation to furnish it. We held our first management college class there in October of 1979, and then did one in November, and December. These were four-and-a-half-day classes, crude by our later standards, but effective. The November class was made up of 18 IBM quality managers who thought they had no problems. However, they participated well and realized that most of their difficulties originated with each other. The company was vertically integrated in many respects so they were sending their problems along the line. At the end of the week we began a relationship with that company that lasted for several years.

It didn't take long for us to realize that we needed our own dedicated classrooms and buildings. Philip, Jr., found a building at the corner of New York and Canton Avenues just off Park Avenue, in Winter Park. We were able to obtain 1600 square feet, which was enough for a classroom called the "Gold Suite," plus some office space. After years of dealing casually with buildings and space in terms of millions of square feet, it was another feeling entirely to be agreeing to pay for something for a fixed period of time. Eventually we had the whole building, all 5000 square feet of it.

I invited two people from ITT to come and join me. The corporate staff in New York and Brussels was being sliced up and everyone was willing to look out for other opportunities. Bob Vincent was preparing to retire from ITT and would be coming down in the spring of 1980. He would become the Dean of the College. We obtained the services of a couple of part-time secretaries, and everyone was treated as an independent contractor until we found out how to deal with Social Security, health-care plans, and such. All of that happened on the first of January 1980, when we had a real company with real obligations to its employees. I still was not taking any salary, but there was hope. The classroom was scheduled out for several months with a class every other week.

Along this time I was introduced to a lawyer, Bill Grimm, who was part of a new firm downtown. I showed Bill my plan for PCA, which included projections that the company would grow to where it could go public in mid-decade. My estimates and plan didn't mesh with his experience, but he decided I was serious about it and became our attorney. He guided us through a dozen years of originality. In the same manner we found an insurance agent, a pension and health plan specialist, and several others who stuck with us through the years. The Lord always provided just the right person for PCA at all levels of the company.

To support the College we started what we called a notebook factory, with our little Savin copier, to put together the material that students would receive. On Friday afternoon, after the class had left, we would all take off our shoes, turn on a radio, and fill the new notebooks. The empty binders would be laid out at the student's table and then we would take the contents, one page at a time, and form a line walking around the room. It was fun and memorable. We learned about the "white boards" which we would use to write on in class. There were markers which you could erase, and there were markers which did not erase.

One of the advantages of our location was that we could walk our students to lunch on Park Avenue. This let them get exercise and drink in a little local character. Two restaurants wanted me to pay in advance when the reservation was made for the 25 lunches since they had never heard of us. Three others welcomed us with open arms, and over the years we essentially paid their complete overhead with our scheduled lunches. In return, they listened to us concerning organizing and planning the meals; with this cooperation we were able to produce Zero Defect lunches on schedule. We never did business with the ones who turned us down, at least not until they changed management later on. Before much time passed we had a full-time employee who just made arrangements for lunches and meetings. We were as careful to not take advantage of the restaurants as they were careful to keep us happy.

The students (executives and managers mostly) came to the Quality College with the idea that if we were going to teach them that an operation could be run properly, then we should be a living example of just that. They poked into corners, interrogated the staff, and expected things to happen on time. We also took that objective very seriously. Each associate was selected for employment based on their desire to get everything done right the first time. All their PCA training was aimed at enforcing this attitude and helping them learn to implement it. When something went wrong, we immediately took corrective action and explained to the class exactly what had happened.

Each Friday we had a graduation lunch for the Management College and told them we couldn't remember having a better class. And they were the best; it got better each week. For the first two years we would have a buffet at my home on Monday evening for the class so everyone could get to know one another and relax a little. It worked out well because they all wanted to get back to the hotel by nine o'clock in order to watch football.

My first priority for the company was that each associate would be carefully selected as a result of her or his personal commitment to really contribute and that we would all treat one another, as well as the clients, like ladies and gentlemen. Senior people in business just weren't very nice to the lower-level employees—at least that had been my experience. I conducted orientation programs for all employees as they arrived, and we set up a "Family Council" system as soon as there were about 10 of us. Each month all of the associates would gather together for an hour or so. The status of the company was reviewed, special reports were made, questions were answered, rumors were discussed, and everyone had the chance to make whatever comments they wished. It was always a struggle to get the professionals to attend each and every one of these, but I insisted. We had over 120 of these meetings before I retired.

All of the associates really appreciated these sessions. After each Family Council I thought about the years spent in companies where management was so distant and untouchable. Standing in front of those people, seeing their trusting, expectant faces, and realizing that they were giving their all to the company made me realize each time that this was the right thing to do. It is a very rare management that will voluntarily take the time and effort to conduct such communications. Whenever they do even one, such as a "ZD celebration," they really have a good time. They talk about it for months. Why they don't do it regularly is something I do not really understand. The common thread of refusal usually seems to be based in the Human Resources Department. They are always afraid of starting a precedent or something. I think they are afraid of the troops.

We never had a dress code as such but everyone dressed in a manner that was courteous to their coworkers and gave a good impression to the clients. We did not tell the students what to wear but most of them wore jackets and open shirts. In early 1980 one of the tenants in our building failed and the area became available; it was big enough for a 14-person classroom, which fit our new Executive College perfectly. We had determined that senior management did not need to know much about implementation; they had to understand the concepts. So we began offering a two-and-one-half-day course for senior people, and it was booked quickly for the rest of the year. Phylis (my daughter) had just graduated from Rollins College and came to work on a temporary basis, until she could "find a real job." We asked her to design and stuff the new room, called the "Blue Suite." She did a great job and we used it for a long time.

I selected people to be instructors based on their experience in real-life management and excluded anyone with a quality-control background. We also avoided professors or those who had been teaching anything. This way we could teach them to instruct properly. The certification period usually took several months but they learned quickly and worked hard on it. We began to discover the need for filming some of the material in order to assure consistency. Also not every person is so interesting they can be listened to for hours. I had been opening each class, which took the first morning, but now that we would have parallel sessions we began to teach the other instructors how to do that module also. One of the cases I wrote was the "Wheelbarrow Company." We would ask the class to rate their own company on the Management Grid contained in *Quality Is Free*. Then we would pass those ratings out the door to the college secretary.

I would use the white board to describe the wheelbarrow they had planned to make and then relate how a few compromises here and there turned it into a financial and organizational disaster. Then students would take the roles of the seven officers of the company and read their comments about the situation. At the end they would evaluate the Wheelbarrow Company on the Grid. By this time the evaluation of their own companies would have been compiled by the college secretary and slipped back under the classroom door. The two were compared by being displayed on the viewgraph screen.

Classes always rated the Wheelbarrow Company at the lowest level on the Grid, in sharp contrast to their own evaluations. We carried this further by evaluating the comments that had been attributed to top management in the story and showing how much money they were spending by not following their own procedures and requirements. As a result the students began to get the idea that they knew more about the Wheelbarrow Company than they did about their own. It was always an awakening to them and set up the rest of the course so they could learn how to make changes in their corporate culture.

The Wheelbarrow case had become cumbersome because I was the only one comfortable in presenting it. For that reason we decided to put it on film. Phylis found a blue wheelbarrow in a hardware store and we found a company who had one of those new-fangled video cameras. It was necessary to hire local actors to do the parts of the executives, and I described the original wheelbarrow design up front. As it all worked out well, we began to realize that this was a good way of doing things for us.

Doug Hoelscher of Tennant and I had gone to Japan early in 1980 to visit his operations and see what was happening in other areas. (Doug, Roger Hale, and Ron Kowal wrote a book *Quest for Quality* in 1989 which described their awakening and progress. I recommend it.) As in the 1975 trip, I was surprised to be recognized immediately over there and treated to much respect and inquisition. They were well versed in my writings and were in the process of translating *Quality Is Free*. Some of my notes of 20 years ago of what my Japanese friends told me:

- They have misconceptions about the United States and American management, know nothing of European or U.S. history. (About as much as we know of Japanese or Asian history.)

- There is said to be no word for "love" in Japanese language. Obligation and duty are used instead. The main objective is to work; the family is not so important; it is macho to stay out late with fellow executives.

- Be arrogant, don't show any weaknesses. Don't take vacations as a matter of pride. There is an extremely low absentee rate; if sick, they will usually call in to take vacation.

- All people in the company are supposed to be equal in terms of membership in the organization. Respect for each job is the key to all this. However, my observation is that some are much more equal than others.

- The Japanese Union of Scientists and Engineers (JUSE), which has been the group promoting Dr. Deming's work, is interested in doing something with me. So is the Japan Management Association (JMA). It is hard to figure out exactly what they have in mind. We will try to work something out with the JMA in partnership with PCA. JUSE is not interested in management and the JMA are not very good promoters.

- Dr. Ishakawa (the true leader of Japan's quality revolution) runs 6 four-day courses for JUSE each year for Japanese top management. It is heavy in statistics, but they all pretend to understand and use them. Suffering is a mark of manhood in this culture. Ishakawa is patient about getting his own way. He has made everything happen here that has happened.

- People in the manufacturing plants are different from the ones I have met with the business suits; they are down-to-earth and very nice to be with. They do quality as I describe it: figure out the requirements; train everyone to understand them; and then get the job done right the first time. They have learned to live the complete system of quality; Americans are always looking for some technique to make it work.

- One Japanese quality manager asked me about Mil-Q-9858 (which is the U.S. Department of Defense quality specification) and said that American managers felt it was too tight. I said that I thought it was loose and forgiving. He said that was a new thought for my country. We agreed on that.

- They hate the number "4" like Americans avoid "13" and love Johnny Walker Black scotch. At least that is what they tell me.

- When I spoke to the Quality Control Society they were very critical of American quality and of the quality-control professionals. They thought I must be very frustrated. I told them that they should not underestimate the Americans, that they were sometimes slow to get interested in what was good for them, but that they would be turned around on quality in a few years. They thought I was being polite.

When I returned home, IBM–Raleigh, led by General Manager Alan Krowe, was beginning a serious quality improvement process and asked for some help in putting the training package together. Alan was the first senior IBM operating executive to start to change the corporate mind about quality. We made a 20-minute film of me walking around the Quality College talking about the Four Absolutes. Soon all of our clients wanted one of those for themselves, so we developed a compact way of doing it. Park Avenue in Winter Park had many nice stores whose owners were very cooperative with us, so we used their facilities as locations. I would start out in my office talking about quality in general and how important it was to the company. We would shoot the same scene several times using a different company name in each shot. We would then go to the parking lot, an art gallery, a men's clothing store, and a restaurant, and we would cover the Four Absolutes of Quality Management in the process. The client company's name would appear three times in their film. This way we could charge them a minimum amount and still make a little money on it. I began to realize that we were going to have to supply clients with in-house training material one of these days.

Classroom and office space was getting tight; we arranged to obtain the fourth floor of a building across the street. That let us build one more large classroom (the "Silver Suite"), and three smaller ones. We were growing very fast but doing it out of our cash flow. It was getting hard to fill the demands of our clients. The results they were getting led them to refer others, particularly their suppliers, to us. I was busy making speeches, and writing new classroom material. At this point in time I was the Creative Department, but Phylis began to put together some illustrators and other product people so she took it over. We were able to standardize on viewgraphs, brochures, and such. It was nice to be able to have material everyone admired. Philip, Jr., was working on trying to get us some word processing and computing capability. We often forget how primitive office equipment was in those days, not so long ago.

In September of 1980 we went back to Greenbrier for our physicals and Dr. Morehouse reported that there had been a significant change in my cardiogram during the stress test. In those days the test consisted of running up and down a short flight of stairs. He suggested that I lose some weight and become more active exercising. When we returned to Winter Park I went to my family doctor with the report and he sent me to Dr. Burton, a cardiologist. He put me on a treadmill, hooked up to all the equipment, and took me back off after about 30 seconds. The upshot was that I went to the hospital and had a heart catheterization examination, followed by an operation in which five bypasses were made of the coronary arteries. In between the two events, he let me go back to the office for long enough to tell the staff what was happening.

The courses were laid out well by that time, and I was not needed for teaching. However, the company was still fragile and needed a gentle hand of leadership. It could continue to survive and prosper without me, but it probably would not grow. There were several people in the organization who could manage the company and another few who would destroy it by their management style. I wanted to make certain that the family knew the difference. I wrote a set of instructions, on my typewriter, suggesting a very specific set of actions for the company if something happened to me. I also noted who should not be given positions of authority and gave the family a choice of two people who could become CEO. The letter was sealed and given to Philip with instructions to leave it that way unless I didn't make it. Then all the staff, and the students who happened to be there at that time, gathered around me and we had a prayer. The surgery and the recovery went very well; I became a stronger, and healthier, person because of it. It gave me a renewed interest in wellness. I taught a class five weeks after the surgery. Philip returned the unopened envelope and we shredded the letter.

We added another location, called the PCA building, without much imagination, that took care of all our personnel needs. It contained 32,000 square feet of space. Phylis was busy with getting furniture for the offices, I was busy demanding that people keep their offices clean, and Philip was busy figuring how to pay for everything. Keeping offices clean is not a natural action. It is a mark of honor for many professional people to have papers piled up on their credenzas and desks, but it looked bad to the clients. I would tour the operations every day, at least once, to see people and show the flag. I would leave a chocolate Hershey's Kiss on a clean desk and nothing on a messy one. They got the idea quickly. People didn't eat the kisses; they saved them. I would eat them on the next visit until I realized that this was not a good idea for my figure.

We were getting a steady stream of companies wanting to come and talk about their strategy. The business was building steadily, and we wanted to reach out to small companies also. In 1979 total revenue was around $400,000; in 1980 it was closer to $2 million, 1981 would be $4.5 million, and 1982 was planned for $8 million.

Bill Sabin, my editor at McGraw-Hill, suggested that I take another look at *Sweet Way* in the light of more emphasis on family management. I liked the idea and went to work on it. My portable typewriter still weighed 21 pounds but I dragged it along on my trips to use in flight and at night. Also I had a new electric typewriter at my desk (a habit which never failed to embarrass other executives), with a new feature that let you see what was going to be typed before it was printed. My grandson, Charlie, and I posed in the computer room for the picture on the back cover of the new *Sweet Way*. The new edition was well received, which made me begin to think about a new book on quality management. It would take a couple of years to get around to it but I began to think about it. In the meantime, I kept writing short pieces called "reflections" which we printed up and sent to clients.

The professionals were working hard to become certified to teach in the college. Bob Vincent had broken the material down into modules and as they learned each one he approved one. I wanted each student to hear and see the same things. For that reason the best instructors were those who had no real quality experience. We wanted ones who had been in management and were tempered by real-life success and failure. The conventional wisdom of quality control and such was really bad news, as it was aimed at detection; we taught prevention. As I would develop a new module, Bob would watch me teach it, document it, and then bring the others on board. Soon he would not let me do that one any more and that brought forth a different module. After a while he only let me do a weekly question-and-answer session with the classes, and said the modules kept changing under the guise of improvement. Actually, I would teach whatever I was interested in that day, which did not lead to consistency.

Qualifying as an instructor was a difficult process. It was necessary to learn the material, of course, but presentation was equally important. John Monaghan, of my ITT days, came down to help us in this area. He was able to show us how to avoid the "noise in the channel" irritating actions that disturb the students' concentration. He helped organize the material into a more logical sequence, and he showed the instructors how to show viewgraphs. Those attending the College often asked if there was any way for them to learn all these things. As a result, we included a half-day "communications" module designed by John. After all, they had to go back to their companies and explain what was happening. Some of the more arrogant companies looked down their noses at actually being taught communications. Once in a while there was a revolt, so I would invite those who knew so much about the subject to stand up and tell us what should be done in three minutes. They always flubbed the opportunity and class went on better than ever.

The associates were putting forth more effort than they had ever emitted at any other company. This always worried me because home life might get ignored, and those who do the real work in the office might not be appreciated. For that reason I established the "Beacon of Quality" award for all hands, and the instructor certification plaque for that group. All associates, except the chairman, were eligible for the Beacon award. We just provided a list with everyone's name on it and people voted. That first year three people were selected and the awards were presented at our company "picnic." This was an annual black-tie dinner dance with spouses. It received that name because when I suggested the dinner, my son, the comptroller, said we couldn't afford that—why didn't we just have a picnic? So we had a dinner and called it a picnic. Each year we had a real picnic anyway, at Sea World, or perhaps Disney World. Everyone would bring their kids, and it was a wonderful time. These things did not cost much but reaped eternal rewards. People appreciated the company taking them seriously.

After the certification awards for instructors were given at the "picnic," a local furrier, Art Labellman, came into the room wheeling racks of fur coats. We asked the wives of the newly certified to come up and select the coat of their choice. They were dumbfounded but pleased. The giving of the coat was explained by my thought that the wives had given up their husbands for a while and we thought it only fair that they be compensated.

When we certified our first female instructors we gave their husbands a blazer with PCA buttons.

As we began to leave 1981, we developed a special program for smaller corporations. Analysis had shown that this was going to be a big market and of course there were millions of these companies around. It didn't take a high level of mathematics to show that almost anything multiplied by those numbers would make money. We also decided to begin making videos that became the "Quality Education System" (QES). Clients were complaining that it was difficult to teach the employees about their role without some material. This was reasonable so we had a brainstorming session and arrived at a list of 15 subjects that would need to be discussed. The idea was that we would have a 15-minute tape on each subject, a lecture led by viewgraphs and a trained facilitator, and then a discussion of how that particular subject applies to them in that company.

That weekend we went down to our getaway house, and while there I wrote all 15 scripts. They just poured out. We wound up using all of them pretty much as they were. Some changes were made because we writers just cannot keep our hands off the material. We set up a team to produce the support material and made plans to film the scripts. This put us in a whole new business of hard products. It also showed clearly that we had been concentrating on the wrong end of the telescope. I thought the managers and executives we taught would go back and teach. Some did but mostly they just couldn't handle it. It was going to cost about $600,000 to make QES, but we had revenues of $1,000,000 a month in early 1982 and even though there was a recession in progress it didn't seem to affect us. We were in the process of opening six new classrooms in a new office park about six miles away from the current offices. We also hired five new instructor consultants and had begun their training period. Our executive committee was very bullish about the way things were going. There was no competition in quality management except for a few individual operators, and our clients were achieving real success. More importantly, they were bragging about it to their friends in other companies.

We had established a line of credit at Barnett Bank, using the PCA building, my house, and my personal guarantee as security. The Barnett people really went out of their way to set our folks up with checking accounts, pay depositing, and personal loans. The company had a good cash flow and money in the bank. Furnishing the Maitland offices was a big expense, paying for QES filming added to it, and we had a dozen people in training who were producing no revenue.

Around June the recession began to hit corporations and, true to form, they immediately began to reduce what they were spending on education and training. Our revenue dropped from the million a month to half of that, in an instant. We drew down our line of credit to take care of commitments, and laid out an action plan. I didn't want to lay off people since we had spent so much time and money training them. We were able to reduce expenses in several ways, but needed to cut compensation somehow. We called a Family Council meeting and laid it all out. I figured we needed to hold out until December or so. My suggestion was that instead of laying off 20 percent of the people, we would all take a 20 percent pay cut. I would take zero salary. They agreed right away, and we prayed about it. One of the traditions I set up at PCA when it began was to start each meeting with a word of prayer. Sometimes when I was not at a meeting they would forget, but overall it was something that helped calm the company and aim it in the right direction.

Our friends at the bank went ballistic when our revenues began to fall. They shut down my personal line of credit and began to make noises about wanting the company's line paid off in full, right now. We had assets that were of much higher value than our loan even though they were real estate and equipment. That did not impress them, of course. Assets are worth nothing if you can't sell them to someone. I did some of my most creative writing in providing a weekly status report of our progress in improving revenue and reducing expenses. It was important to not let anyone think we were in a terminal state; that would have shown through to the clients immediately. We kept getting new ones but we didn't have much to sell the ones who were finishing their management education. Much of the problem came about because companies like IBM, Milliken, Stevens, and others had put all the folks through Quality College that were planned. If we had begun QES earlier, we would have been in good shape.

The bank gave us an "adviser" who immediately suggested that we stop traveling, and gave me a lecture on how "marketing types," like me, were impractical when it came to money. How consultants are supposed to deal with clients without traveling was lost on him. We got paid for spending time in our client's operations; it was a main source of revenue. I kept trying to sell the PCA building which was worth about two million dollars, but the whole thing about recessions is that no one wants to buy anything.

Each day I made the rounds of the company seeing how contacts were going, and every Friday morning we had a Chairman's Chat in the lounge. We picked up several more clients and were beginning to work with them. Unfortunately, it takes several months before a real cash flow leads to recovery. The bank executives called me regularly and we had a meeting twice a month in which they revealed their complete lack of understanding about the world of business. I kept asking why they were so panicky when we were paying our interest on time. Of course, the auditors had placed our note in the questionable category and they did not want our blood on their record.

Philip, Jr., had the tough part, paying our bills. We could do well if we didn't have to pay for anything. We wrote a letter to each major supplier saying that if they could just charge us for their costs at present we would make up the difference after the first of the year. We had established relationships with most of them that provided a positive response. Philip and I went to see several suppliers and they assured us of their support. Two or three would have no part of it, so we paid them in full and in keeping with my policy they have never been invited back. They made several attempts to reconcile without success.

One recurring problem came in paying for QES. The filming production costs were hard to meet. The notebooks and graphics were being done by our own people and we were paying them anyway. We learned to not waste much time doing the videos. The actors all came from the local semiprofessional theater and most could learn their lines quickly. We filmed them all on location, with the cooperation of those in the area. Everyone likes show business. I would introduce each segment, usually by talking to the actors in their roles. Then the point of the session would be dramatized. One was about the Flypaper Company who spend most of their money going around the world looking for flies who were weaker than the glue they were making. It was an original way of teaching quality management; the segments were amusing as well as interesting. I thought everyone would be thrilled by them, but was concerned to learn that the consultants and instructors were not happy about QES. Apparently they thought that films would replace them. At any rate, they were always suggesting that we postpone or eliminate the series. They did not see much of a market for hard material. Since there were hundreds of millions of people who needed to understand quality management, it seemed to me that this was an erroneous assumption. In fact, it really ticked me off and I called everyone together to talk about the future of the company.

I said that we would always have a Quality College that taught managers, executives, and client instructors in person but that could only take care of about 3 percent of a client company's personnel. Also it was not a very profitable business. We would have consulting activities with clients but even though we charged them well for that it was not a good way to produce revenue. It was heavy in people costs, and clients really didn't like to have anyone poking around their operations. The future was, to me, in hard products. If we could provide video- and audiotapes, that would give the clients the education and direction they needed, then they could do it all at a much lower cost. After the development was paid for, videos were about 90 percent profit. Since I wrote most of the material, the only cost was production.

People have their own agenda and most of those schedules are not too productive. I let everyone know that we were in a survival mode and that while all thoughts and ideas were welcome, we were not doing consensus management. I would make the decisions and we would do it that way. If anyone wanted to put their house and provide their personal guarantee on the bank notes, I would give them an extra listen. I reminded them that we had given everyone stock in the company and that if we were to ever go public with the stock we had to offer more than a bunch of people standing up in front of another bunch of people. We needed a company, not a gathering. One or two did not buy all this so I encouraged them to go start their own companies. They left, but never did anything.

Several times each month I went off to make a speech somewhere to generate revenue and to get a feel for what was happening in the world. I visited clients and was pleased to see that they were all doing well. It was tight but we were going to make it; the opposite never occurred to me.

In 1979 revenues were $400,000 with 5 associates

In 1980 revenues were $1,664,000 with 16 associates

In 1981 revenues were $5,333,000 with 67 associates

In 1982 revenues were $7,897,000 with 94 associates

(Eighty percent of 1982's revenues were in the first six months)

PCA—THE EXPERIENCES

We finished production of the Quality Education System (QES) and began to present it to existing clients. It consisted of 15 videotapes covering the concepts of quality management, individual notebooks, viewgraphs for the facilitators, and an instructor's notebook. Since QES represented a new concept in quality education, the clients did not absorb the idea immediately. It took a while for most of them to realize that we were offering them the opportunity to be the power behind their process. They could do the "teaching," and the material would assure that everyone learned the same thing.

Our instructors did not like to tell their classes about it because they considered that to be "selling." How they thought our clients would know that we had gone to all this trouble to provide them with an education capability, I'll never know. When you have something everyone wants it is only necessary to expose them to it; there is no selling. Professionals become self-centered in every business. Once clients realized what it was, QES became a big hit even though it took several months to begin to produce revenue. We insisted that those who would be facilitators inside their companies come to school for a week, after they had completed the Management College. Then they could lead the students through the material using examples from their own organization. We planned to lease the tapes and sell the workbooks. The idea was that each person should have his or her own notebook to mark up and keep forever. We found that some of the training people in client companies would duplicate the notebooks or cover the pages with plastic and use them over again. Much of this was killed when I explained to their CEOs how cheap and uncommitted this made them look to the employees.

We were still struggling with the Barnett Bank people who were very worried about their money, although we had not missed an interest payment. Then two interesting things happened. A local individual made us an offer on the PCA building which I accepted. The mortgage company he selected came and approved everything; then he decided that he was in a position to beat us out of about 10 percent of the deal. I canceled the whole thing right there. "Neither a Scrooge nor a patsy be," as we say in West Virginia. The mortgage company representative said that they would finance the building for me personally but would not finance PCA as the owner. All the cash flow I had personally came from PCA, from whom I was not drawing a salary, and my net worth was close to zero. Although PCA's revenue was picking up rapidly, we were not out of the woods. Somehow the mortgage company's determination didn't make sense, but I eagerly agreed to do it even though I didn't have the $150,000 necessary for a down payment. The Lord would work that out, I knew.

The money the company received from my purchasing the building paid off Barnett, and I became a landlord of the company. At this time the president of another bank, a smaller one, invited me to lunch. He understood our situation and offered to finance our company as we grew. He also made me a personal loan that let me handle the building deal. So as 1983 dawned we were again financially free and bound to never to be in debt again. I went over to a golfing buddy's car shop and bought myself a new Jaguar. I also asked Philip to put me back on the payroll. As soon the increased revenues made us solvent, I gave our associates back the money they had lost in the pay cut earlier the previous year. Philip decided that he had enough of being comptroller for a while, so we hired a new one and he took over administration. Neither he nor Phylis and I had any problems working together. We did not compete; they didn't write books or teach, and I didn't do what they were handling.

We were soon involved in a surge of business. Our clients were recommending us to their suppliers and friends. We had a very imposing list of clients and were getting a lot of media attention. General Motors asked us to work with them. Jim MacDonald, their president, and I talked about the idea. Jim had been at Pontiac back in the 1960s when I got to know their people. He was a very quality-oriented executive and even managed to smile when I picked him up in my Jaguar during his time in class.

I suggested that working with GM could be a problem for us because we would have to build up our staff in order to handle them and they might change their mind suddenly. They offered to purchase an equity position in our company in order to make things more permanent. Jim and I agreed on the price ($4 million for 10 percent with an option for another 10 percent) right then. It took the lawyers several months to work it out but in the meantime we worked together. (The net result of their purchase was that they made money on the investment, even after paying our fees for several years, when we went public in 1985.)

GM was a good partner, sent all its executives to school, and had us set up a manager's Quality College for their people in Michigan. I always remembered Woody Allen's comment that "when the lion and the lamb lie down together, the lamb doesn't get much sleep." They never gave us any trouble as far as their ownership went. Also we never gave them any discounts; they got exactly the same treatment as other clients, and they insisted that it be that way.

The stock GM received came from the principals of the company. It was very nice for us to have some real money for a change. I was able to pay off my condo and car, and for the first time in my life I was debt-free personally. The rest of the family and some of the key executives divided the money. The company had no need for it now that we had learned to manage without debt. Our cash flow was very positive and we were using it to grow. It is harder to manage success than tribulation; people don't listen well when it comes to dividing the spoils, but they do when we are launching lifeboats.

We had meeting day at PCA once a month and each member of the executive committee was required to attend. Meetings were scheduled for all day on that Monday and Family Council was held on Friday of the same week. The operations review system was based on my ITT experience, which demanded that everyone look at all the numbers, listen to all the problems, consider all the possibilities, and help make decisions. Everything was dealt with and, as in ITT, there was hell to pay only if one had not been open and honest about something. The meetings opened with a prayer and everyone was nice to each other. We fought the problems, not the people. This tone was evident when I was around and often disappeared when I was not. People do have their own agenda. Usually they do not see the future past their own commitments. No one looks at the whole organization except the CEO.

One time the European contingent explained a revenue result below plan by noting that Easter had caused people to not want to do business. They spent the next 20 minutes explaining to me why the occurrence of Easter had been such a surprise to them. We even got out an encyclopedia and noted that we were able to know exactly when Easter was going to occur for each year in the next century, which should be plenty. We remembered that the purpose of the company was to make our customers, suppliers, and employees successful. It was also not to make trouble for each other. Business should be fun, and we should be the absolute best in the business. We had to be unselfish in our dealings.

A BBC producer, Brian Davies, called and asked if he could spend some time with us. Brian had produced several notable shows for BBC, one of them a study of lilies worldwide. I was struck by his sensitivity in that piece. After showing the massive and dedicated places of the world for raising lilies he filmed some lilies growing wild on a rocky slope in Greece. They were the most beautiful of all because, as the narrator said, "They chose to grow there." Brian came and lived with us for a month. He went everywhere I went, attended family and company parties, listened to me teach and speak. Then he returned to London and wrote us a proposal that we would do a 40-minute show called *The*

Quality Man. He wanted to do it when my wife Peggy and I took a trip to Scotland that June. I said we could work it out if they could help me get on the Murfield golf course near Edinburgh. It was an impossible thing for an individual to arrange; The Royal Company of Murfield Golfers was a notoriously difficult organization. When Tom Watson won the Open at Murfield a few years back, he and Ben Crenshaw went out on the course later that same evening, after the crowds had all gone home, to play a few holes just for relaxation. The club manager came out and ordered them off the course because they were not members.

Somehow BBC did arrange it, paying a fee in the process, and we drove down from Gleneagles where we had been staying. An associate and his wife met us there and the four of us set out to play a few holes with the cameras and microphone picking up the conversations and action. Women weren't permitted in the clubhouse, of course. However, the manager did unlock a toilet out on the course for the women. When he had enough course shots, Brian set up in a little valley that provided a view of the Firth of Forth and planted me with my back to the view. Then for two days he asked me questions and filmed the responses. His monthlong visit had taught him more about my philosophy and stories than my staff had picked up. The result was a really fine piece of work, interesting and informative. They have sold hundreds of the tapes and shown it on BBC several times. They have made thousands of dollars on it; I got $300 for my efforts. However, it has been a marvelous help in building a business and reputation.

I began to work on a book to explain quality management in a much broader sense in terms of management action required. My experience as a consultant and teacher for the past four years had shown that management really didn't like their employees very much. They continually hassled them with action that kept the pot boiling. They honestly didn't seem to understand that the people were their primary asset. Competitors all around the world could purchase the same equipment and material; the only advantage a company could have would be its work force. My personal experience working at low levels had shown me that management was insensitive and that they assumed people down the line were not too bright. The title of the book as I planned it was *The Art of Hassle-Free Management*. However, the publisher hated it, which is fairly normal. But they fought a bigger battle than usual, so somewhat in jest I suggested *Quality Without Tears* (QWT). They liked that and my favorite phrase, the original main title, became the subtitle. All of my books after *Sweet Way* had the "art" of doing something as their subtitle. The main emphasis of QWT was to spell out the concepts in clearer detail, the "Four Absolutes" as I called them. If one could understand these, then the whole business of quality would be revealed. Management is so technique-oriented that it is hard to get ideas through to them. The story chapters included one on a big corporation's wasted planning effort, and a thinly veiled take-off on Dickens's favorite story, "A Christmas Carol." I changed it to "The Quality Carol." In my story Marley took Scrooge to a warehouse to show that he was condemned to fix all the shoddy products he had shipped out of his factory during his work life. In the next bin a lady who had been in charge of luggage for a large airline spent eternity trying to unite luggage owners with all their bags she had lost.

When the book came out, the response to "The Quality Carol" chapter was so strong that I decided to make a training movie out of it. We were able to get Ephram Zimbalist, Jr., to play the lead and a fine group of East Coast professionals for the supporting roles. We built a set in Cocoa Beach and spent a fun week shooting the movie. Our star proved to be a fascinating person and a complete professional. He was always ready, knew his lines, and provided a lot of support for the rest of us amateurs. The film was well received, primarily because the production was so good. But I had to note that although people enjoyed it, and bought it, they did not seem to take the message personally.

For the first time in my work life I was able to write and think during the day. PCA ran on a day-to-day basis in a very calm manner. We had thought out the procedures and policies and incorporated them into a strategy that everyone could understand. There were no sudden changes; everything was planned. So the panic calls and noncompleted assignments usual in most companies did not exist in PCA. Things were going so well that I was concerned that the associates would forget to appreciate all the wonderful things that had happened to us. For this reason I established a "Thanksgiving Week" that would occur in April each year. The theme was to thank all of those who had made it possible for all of us to be living so well. The quality improvement team took charge of the event. There was something happening each day of the week.

On Monday morning the employees were thanked. Each of the associates received a memento when they came to work, and the committee came by the monthly operations review and thanked the executive team, usually by giving each of them a rose or something similar. The outside directors were thanked at the board meeting that day.

On Tuesday the suppliers were thanked. They received letters and in some cases a personal visit.

On Wednesday we had a prayer breakfast to thank the Lord. A guest speaker was invited and most of the associates came to the session which was held out by the lake at 8 a.m.

On Thursday we thanked the clients. Many were called, many received letters, and those who were there in class each received a personal greeting and a memento. Thursday was the day, each week, when I usually did an open discussion with the classes. This consisted of an hour in the auditorium, and we taped the sessions for distribution and sale later.

On Friday we thanked the families. After work they were invited to the building for a buffet and some fun and games. One year the kids threw baseballs to dunk their parents in a tank of water; another year there was a ski show. (We also had the real picnic in the summer.)

On Saturday we had the Thanksgiving Ball, our annual black-tie affair where we presented the Beacon of Quality awards and thanked the organization in general. We had a great band, Marshall Grant from Palm Beach, and everyone danced, even the young folks who were shy about it at first. Thanksgiving Week didn't cost much, but it did a good job of keeping the tradition of the company alive. New employees do not know the trials of the past; they think everything has always been as it is. Every organization needs an annual reminder of what they are all about.

I had tried several times to set up a Quality College bookstore. My thought was to have a place where students and others could buy my books, of course, but also where they could get golf shirts, pens, and other memorabilia with our name on it. The several attempts to do this through the College personnel met with failure; they just could not seem to get the idea even though I personally took them to the Rollins College bookstore and sent them articles on how the major league teams made money selling logo material. It was viewed as just another of the chairman's quaint ideas. So one day I asked Phylis to assume responsibility for it and took it all away from the others. Within a few months she had a catalog out and a flow of nice material going through our shop. Within a year it was doing over $200,000 worth of business, staffed by a part-time attendant. The students were able to get to it between classes, and she advertised in our magazine which went out quarterly to the mailing list. My thought was that executives all around the world would see others playing golf or relaxing with nice shirts and hats bearing the PCA or Quality College logos. To me that was advertising and marketing; to many others it was just a diversion. Those who had not lived the life of a corporate executive were not able to market to them. They just did not understand what was important. It is very difficult to teach people about things they have not experienced. That is probably why marriage counseling is not all that successful. It also made me wonder about all those who were conducting seminars on quality management who had never actually managed any quality.

When we began to have operations overseas in 1981 it was only a matter of time before those offices needed to be permanent. Clients did not appreciate it when we landed some people and boxes, conducted several classes, and went back home. They wanted on-site support; at least they said they did. We set up offices in London and Brussels with plans to do the same in France and Germany. As we hired people, we brought them to Winter Park for orientation and training. The clerical staff would come for two weeks, the instructors and consultants for six months. The idea of bringing a secretary across the Atlantic (or from Singapore) seemed like an extravagance to many. However, those people never forgot what they learned at Winter Park, and they built relationships that made them valuable associates. Left to their own, branch offices never do any kind of adequate orientation. It is a very difficult concept to get into the minds of those who see the world in terms of revenues and profit.

We started annual Alumni Conferences in order to help clients bring each other up-to-date. All the speakers were client people talking about their experiences, and the results were wonderful. It was necessary to charge for the conference in order to defray expenses, but each year over 200 people showed up. We would display new products and I would have an open discussion with the group. Around 1985 I could begin to see implementers in the quality management field beginning to search for tools to implement quality management rather than managerial concepts. It was disturbing, and meant to me that a great deal of effort would be wasted in the field. Business was just beginning to get interested in organized drives to improve their quality. What we taught was a culture change which required management commitment, education, and action. What most executives wanted was a series of actions that could be delegated to a functional department like Quality or Human Resources. A wide array of consultants were springing up eager to satisfy this illusion. They would define quality any way the client wanted it; they would not insist on management being educated; and they were willing to do everything for much less money than we were. All of this scared the executive committee. But the key point was that none of these new people had videotapes and other hard material. They did not have the money to compete with us in those areas, and the more material we could create, the better off we were. I directed that we would not respond to requests for bids from potential clients. If they wanted to succeed, they would have to do it the real way. We kept growing, reaching $20 million in revenues in 1985.

One company program that was being well managed was contributions to charities. Originally I had set up for PCA to distribute 10 percent of after-tax earnings to various institutions such as colleges, children's assistance programs, arts, and such. We excluded churches because I felt that the congregations should take care of themselves, and besides that, there were hundreds of churches in Central Florida. We finally settled on 5 percent of earnings, which came to be a tidy sum. In 1985, for instance, we gave away around $300,000 and we didn't tell anyone about it. We had set up a committee early in the company's beginning and had hired a consultant to help us out around 1983. Larry Kennedy was a pastor with an MBA; he had a little operation that helped charitable organizations to manage themselves better. We asked him to make certain that those who were asking for money were legitimate. We said that we wanted to give money to "real people helping real people." It is not as easy as it sounds. Here are some of the organizations we became involved in helping, not only with money but with our associates as volunteers. We searched to find the right ones, and asked our associates for suggestions also.

The House of Hope was started by a schoolteacher, Sara Trollinger, who was concerned with the runaway, throwaway teenaged girls she saw each day. With no money and lots of faith Sara built a shelter for these girls where they could get their act together, catch up on education, and feel some love. She started with just a couple and now has room for 25. The parents, if existing, have to come once a week for counseling. Her success rate in getting the families back together is over 90 percent.

The Orlando Science Center has "hands-on" equipment for kids. They wanted to get the first of the dinosaur exhibits. We gave them $25,000 and that put them into the big time. People came from all over to see these mechanical beasts and found out about the center. We continued to support them for years.

The Orlando Museum. Peggy was on the board and we helped them build a wing.

Christian Service Center. These folks have fed hundreds of homeless each day. We funded much of that.

We bought trucks, funded shelters for the homeless, gave tuition money to those who did not have it, funded a food bank, bought equipment for a college to teach the blind to read, and, in general, helped a lot of people to help real people. Our associates knew of all this and were proud to be part of the company. Now and then someone would indicate that we should share this with the shareholders rather than giving it away. My response was that God had made us what we were for this purpose and that we needed to honor our commitment to Him. If they wanted to do it differently, I suggested that they go start their own company.

If we wanted to gain more market share, we needed to develop more material and that was going to cost money. Also I had given away to the associates and my family about 80 percent of the stock in the company. The family owned less than a third of the organization. I thought we should consider going public. No consulting firm had done that before, and when we were approached by Dean Witter they were cautious. They were concerned that we were like an advertising agency where "the inventory went down the elevator every night at 5 p.m." We showed them that we were a product company. The College classes were conducted by people who had been taught how to do it, not by professors or individuals doing their own thing. All of that was carefully controlled. We also showed that half the revenue and most of the profit came from the hard material. My plan for the future was to reduce the percentage of stand-up teaching to a minimum and concentrate on hard copy. We began the process of going public. I didn't know what complicated was until that began.

"Due diligence" means that everything one says about an organization must be verified so that the person considering a purchase of shares will be told the complete truth. This means everything. For instance, part of the proposal copy referred to the sales of my books. Books are a fluid subject; there are thousands of them out in the world waiting to be actually sold to someone; there are some which have been sold that have not been reported as of that moment. As soon as you put a figure together it is no longer valid because someone sold one in London or something. *Quality Is Free* had sales of around two million copies, counting hardcover, paperback, and translations. Several publishers were involved, and it took quite a bit of doing to arrive at a number everyone could accept. It seemed like a lot of foolishness to me in many respects, but we stuck to the letters of the law, and there were a lot of them. We met with securities analysts, most of whom had little practical experience in the world of business. However, they were good at investing money.

After more effort than one could believe, we were launched on the over-the-counter exchange in October of 1985. The stock opened at 13 and worked its way to 34 a few months later. Peggy and I gave shares of stock to several of our favorite charities, funded a Crosby Scholar's Tuition program at Rollins College where each year a gifted child would receive a commitment for a four-year education, and the Wellness Center at Winter Park Hospital.

My reasons for wanting to go public were really three: first, I thought the company needed that sound base of being able to reach out for funds and never suffer the indignities that were placed on us by our bank; second, the associates and family members who owned stock would now be on their own, and I would not feel responsible for them having stock that was not marketable; and third, I wanted our family to have money so they could be independent. All of the associates now had stock worth real money, and several became millionaires. During the years I had insisted that they learn how to use financial advisors and manage their money. We paid for that but had to drag most of them to the sessions.

I didn't start the company to make money, and perhaps that is why we did. I wound up with several million dollars in cash and stock personally, and gave away about half of it, but we had launched a quality reformation throughout the world. By the end of 1985 PCA had sent almost 12,000 executives and managers through The Quality College. Thousands more attended QES and other classes.

1983 revenues were $12,872,000 with 102 associates

1984 revenues were $20,034,000 with 128 associates

1985 revenues were $34,333,000 with 182 associates

PCA—GROWING

Dealing with people in the most agreeable of environments with the most enlightened of managements is still the most challenging of tasks. One of my basic principles was that management should not harass the employees. That was the subtitle of *Quality Without Tears*. Also I discussed it regularly with all members of supervision and tried, in fact, to have the absolute minimum of supervision for the associates. One day my secretary said that she had just left the restroom where one of the female analysts was crying. After some coaxing I was able to learn that the cause of her distress was her boss, who had been yelling at her. When she came out of the room I asked her to sit with my secretary and me in order tell us the story. After she was calmed, she mentioned that her manager was verbally abusive to all of those who worked for him. I went into the next room and called the personnel manager (we finally had to get a department to handle that stuff) and asked about this manager's way of working. I was told that he had been talked to about this and had promised to reform but was internally convinced that this was the way to manage. When everyone left I called this manager to the conference room and asked for his side. He rolled his eyes and said that these young women were overly sensitive. I pointed out that this "young woman" had an MBA, a husband, two kids, and was approaching 35 years old. She deserved to be treated as a grown-up. He said he would try to do better. I suggested that his attempt would have to be made somewhere else, summoned the Human Resources director, and that was the end of his employment. A similar situation occurred with what would now be called sexual harassment when a very senior executive was given his walking papers after several personal discussions failed. These problems did not happen again, as far as I know, but it is hard to get people to treat each other like ladies and gentlemen. If people did, Shakespeare would not have had anything to write about.

My new book *Running Things* was copyrighted early in 1986. It had been written in Savannah the previous year and represented the book I wished I had when PCA began. It told the purpose of an organization and described how to create one. I hoped it would help people think things through rather than just bumble along. Most companies fail because the management effort is incoherent rather than because the company does not have enough money. Hardly anyone believes that.

Writing a book is a fine involvement for me. Everything comes out of my head, so I don't have to plow through other books to lift and footnote quotes from other writers. It is all me. The books on management that I see today are often a compilation of what different people say with the author's opinion placed on top. This wouldn't be bad if they got it right, at least in my case. I can only think of a couple who came close to stating my thoughts correctly. My assumption is that they did the same with others. For that reason I never comment on what others write or do; they can do it better than I.

Running Things could have been called *The Entrepreneur's Guide Book* because it deals with the real-life actions a CEO had to take. PCA was a happy company; the people were proud to work there; it was profitable; the customers liked it because it was so professional. All of this came about because of actions I had taken, or programs I had installed *on purpose*. These actions were not obvious and hardly anyone noticed that there was a pattern or intent. When I explained them to the executive committee or to those who asked, they recognized it but I noticed little adhesion. We had a quality improvement team, for instance, and appointed a new one each year. They did a lot of good work, but their efforts would have been for nothing if the culture was not one that wanted to improve.

Every person who came to the company on a visit remarked within their first half hour that this place was different. Everyone was friendly, to each other, to the suppliers, to the clients; they knew their jobs; they helped each other. All of this happened, as I said, on purpose. The executives and managers who were our students did not recognize at first where all this came from. They thought that it was because the people of Central Florida were just naturally soft spoken and polite. After a day or so of being involved with them they began to see it was a matter of selection and action. We were able to show them how to arrange their place the same way, if they wanted to put forth the thought. This came up often in the coffee room discussion with executives attending the College. They realized quickly that it would require them to pay attention to the people and their environment. Usually this did not fit in with their agenda.

I realized that the company was going to change after I left and also that I was getting ready to be gone. Other thoughts were creeping into my mind, and I was trying to turn things over to others even more than I had in the past. The one rule I imposed was "don't eliminate anything unless you know why it is there; each brick in the wall has a purpose. If you remove it, something must fill its place or the wall will be weakened." This had little effect on anyone; it is hard to learn from someone you know.

My mentality had always been to create something, develop it, show others how to do it, and then move on to something else. Up to this point that strategy had always worked rather well. All the individual functions of PCA ran without problems. Purchasing was dealing with cooperative suppliers, and we ran seminars for them in order to build up relationships. Each year when it was time to do the supplier workshops I would wait to see if anyone began to arrange it. Each year no one would until I suggested it. Then the whole thing would be accomplished in a very professional and effective manner. It was apparent to me that I was becoming part of a problem myself; probably they were waiting for me to suggest events. I began to cut myself out of more and more things in order to give the next generation of management a chance to grow and take charge.

I can't stand conflict, the kind where people are unkind to each other, or yell, and argue. Movies are tough on me; it is hard to sit and watch someone be picked on or beaten up or ignored. The ones where dozens of actors are killed in war or cattle vs. sheep fights don't bother me a bit, but if someone rejects an individual, I get all upset inside. I am ready to talk about any subject as long as anyone wants to discuss it, and dealing with problems, no matter how complex or personal, poses no difficulty to me. But I can't stand conflict. For that reason I went out of my way to make certain that the environment of the company was peaceful and purposeful. Any supervisor or manager who chewed their people out would get a personal lecture from me. Anyone who could not manage like a lady or gentlemen was removed from that post, and usually let go. I fired three people for abusing their responsibility. I can think back over my 40-year career and identify at least 50 people who should never have been put in charge of other people. I can think of only a dozen who were really good executives.

Keeping quality installed in a company is a full-time job for the senior executive. My daily routine, if I was not out traveling, was aimed in the direction of assuring that the proper work environment existed. I wanted to make certain that everyone was proud and productive. Each day would begin with processing my mail which I still stood up to handle, not from respect but because I had learned that it went faster that way. This took at most a half hour, then I would go over to the College and wander around seeing if everything was working. I greeted each individual with a handshake or a hug. Lots of people get very little hugging, so it becomes a sort of a public service. However, it is necessary to learn to wait for them to come to you, not to go to them. This way, if that is not their idea of communication you do not embarrass them or yourself. As part of this touring I never gave anyone directions, but I did do a lot of appreciating for obvious good work. Often the visit to the College was planned to coincide with breaks in the classes, then I could schmooze with the students for a bit. The executive students, in particular, gave me a chance to learn a lot about what was happening out there in the real world. Each of them was very up-to-date on the problems and opportunities in their area of business. We also traveled to the same places and often knew the same people.

Managers knew more about the functions—what was going on in Human Resources, for instance. We never got many from the Quality Control areas but had a lot of new people in the quality field who were trying set up something for their company. I had to be very careful not to keep them past the time class was supposed to begin; if I risked that, the instructor would come over and give me a not-so-gentle hint.

Then I would visit the administrative areas until each associate had at least been waved to. Every person felt like they had a relationship with me, but no one felt that we were buddies. It is dangerous for an executive to get too close to individuals, so close that there is an indication of special treatment. We have to keep what the psychologists call a "social distance." This is being approachable and friendly but never leaving the throne. My opinion is that one is either designed to be able to work this way or not; it is very hard to learn. For me it was natural; my life is full of friendly relationships but very few friends. It wasn't a matter of like or dislike that caused this, more that most people are not very interesting for very long.

Client companies wanted to know about case histories, what had happened to other companies who had implemented the quality concepts and processes we taught them. Just telling them about it didn't seem to satisfy their needs, and articles only whetted their appetites. Actually no one ever does anything much differently after being exposed to a case; they immediately say that it does not apply to their specific situation. But in order to satisfy this need, and also to make our clients proud, we began to film activities in their plants and offices. We called the film series "Quality in the 21st Century." Each film is about 25 minutes long and consists of interviews with people of various levels in that company talking about their activities. Phylis was the master of ceremonies and I did a few commentary moments at the end. We did Bama Pie; Winter Park Hospital; Green Bay Packaging; General Motors; People's Bank; ICI; and a dozen others. It is an extremely helpful series and has recently been made part of an educational project of the Irwin company. The Irwin book *Management, Quality and Competitiveness* was written by Professors Ivancevich, Lorenzi, and Skinner. I wrote 22 "reflections," one for each chapter.

The Quality Awareness Experience film was not going over well in Europe. The middle-class American family we had portrayed appeared to be opulent over there. So we went back to our BBC friends and had the material rewritten into British situations and language. We used the European video format, and while we were at it, added upgraded examples. We did the same thing with QES, dubbing all of them into the various languages involved. My thinking had been that the continent was full of American films, showing in theaters all over town. This didn't seem to bother anyone, so why should training films have to be so European-compatible? Their purpose was to display ideas and instruction, not to provide a mood or a good cry. However, the professional audience, the training and quality-control people, felt that it was not for them, so we gave them what they wanted. The film turned out to be much better than the original, which taught us a lesson in humility. We shot my part in the Sheraton right down the street from Harrods in London. The clients liked the result and the QAE program became popular all over Europe. The material was almost exactly the same as the original, but those ordering it could now say that it was conceived and produced in Europe. Being human, they were not necessarily interested in improving their company; the real concern might have lain more on the way their own efforts would be viewed. The result was a much better product, and of course I had learned once again not to be so certain that I am right about things.

Many PCA associates had children in college, all of whom were looking for summer jobs. It seemed to me that hiring some of them would be good for relationships and would provide us with some useful labor over the vacation months. Most PCA people received three weeks' vacation after two years, and we insisted that they take the full time. This created a need for folks to fill in while they were gone. In 1986 we had a dozen summer interns. The rules were that they had to have real jobs in the company with real responsibilities, and that they would be properly oriented. Over the years we never had any problems with any of these kids and many of them came to work with us after leaving school. One year the human resources director came to say that he had a budget to cover only 9 interns but had 11 applications. I suggested that he decide which two of his fellow associates he was going to disappoint and then go tell them. He found some more budget. The intern program was an example of building goodwill with the employees, at a very low cost, and helping the company at the same time.

When each new associate was being oriented, and we included those client people who came for extended instruction also, they had lunch with me for the ADEPT presentation. I gave each of them a little block of walnut wood with a ball-point pen sticking in a hole. On the top of the block the ADEPT code was engraved. I would give each of them the box containing the block and the pen and then take them through what the words meant:

A is for accurate—we do what we said we were going to do and when we give information we know it is right.

D is for discreet—we do not gossip, and we treat everyone like ladies and gentlemen.

E is for enthusiastic—we try to hire enthusiastic people and then not turn them off.

P is for productive—the more of us there are, the less there is to share.

T is for thrifty—we do everything first class but there is no virtue in throwing money away.

All through the world of PCA, people had these blocks on their desks. They remembered that lunch; they had a photo of themselves with me. This is the kind of activity that I had a hard time convincing most of our client executives to undertake. One of the quality improvement teams turned the ADEPT into a recognition system. Each two months the employees would vote on who was the most accurate, or whichever category was up at that time. They would give out a little plaque at Family Council. This reminder and its ceremonies cost virtually nothing but generated large rewards. This sort of thing is what I think chairpersons are supposed to cause; they are in charge of all the people things.

My primary work effort was making speeches that let me market the company and spread my philosophy. For the most part we charged somewhere between $10,000 and $15,000 for these sessions. However, selected clients or true public service events were done free. In total, I usually did about 75 speeches a year, plus a couple a week at the Quality College. This schedule more than reimbursed the company for my salary and expenses which made me feel that I was earning my money. I really have a problem with executives who make a lot of money and don't do any real work. I can't stand to read the annual *Business Week* issue on executive salaries.

Each speech was unique in itself, and I prepared for it in advance. Usually I worked the Absolutes of Quality Management in using personal experiences to show their application. It was easy to talk about realizing that the word "quality" had no agreed meaning, while at Crosley; to realize that Acceptable Quality Levels meant a commitment to doing things wrong, while at Bendix; to realize that I could set a new management standard with Zero Defects, while at Martin; and to realize that money had to be used to measure quality if executives were ever going to pay any attention to it, like at ITT.

Of course the message was prevention, but it had to be shaped so the audience received it well. People asked me how I could stand to do all these talks on essentially the same subject, and I noted that the audience was different each time. Every actor and speaker figures this out. The key to successful speech making lies only partially with the speaker. Arrangements are a big part of it also, such things as the comfort of the audience, the comfort of the speaker, the understanding by both of what the session is all about. My experience has been that left to themselves, the arranging committee will put the speaker behind a lectern, hemmed in by others sitting on the podium. They will shine a light directly in your eyes and turn the illumination down on the audience. This way you will not see those in front of you or your notes. I need room to roam and despise using a lectern. It is okay for laying some notes on to wander back and review. But the separation from the audience can kill a speech.

In the same manner visual aids are not friendly to the speaker. No matter how well thought out or attractive the viewgraphs are, they detract from the speaker's message. It takes a lot of work to gain the audience's attention and confidence; when that is tossed away, much of the speech goes with it. Let those who deal in numbers rather than ideas use such things. Also the projection machines often malfunction. I much prefer to deal with the audience by myself. When doing a seminar it is a good idea to give the audience individual notebooks. Then the speaker can refer to a certain page number if there is something to share or evaluate. Sending people off in groups to do some workshop or other is not useful as far as acquiring knowledge is concerned. However, it does use up time and lets the people get some exercise while the speaker takes a rest.

All of the travel necessary for such work would not have been possible without PCA's Lear 55 airplane. It not only let me get around the country rapidly, speaking to meetings and visiting clients, it let Peggy go along too. She shopped and visited often while I was doing much of my thing, but attended the talks most of the time. I always think no one is going to come to the speech, and during it I feel that I am losing the audience. But a glance down to her lets me know what is really going on, and she is an excellent critic. When one is the chairman, founder, and chief guru of a company it is not easy to acquire constructive criticism. Even if you do get something useful, it may not relate to the thoughts that are going through the mind at that time.

These thoughts sometimes grow into a largely expanded version. In the book *Quality Is Free*, for instance, I talked about "integrity systems" which described all the disciplines of producing quality. Nine years later this two-sentence thought became a whole book on the subject: *The Eternally Successful Organization* (*ESO*) which was published by McGraw-Hill in 1988. That happened about 18 months after the original idea occurred, which is about par for the book business. It takes most of a year to grind it out after the author puts in a manuscript.

The idea for *ESO* came on a visit to Bethany College in West Virginia. I was speaking to an industrial development council, and as part of that attended a luncheon with some officers of the College. They were talking about how poorly academe was managed and we were all agreeing. However, it soon emerged that Bethany had been there for about 150 years, had no debt, and possessed a very healthy endowment. It began to occur to me that there had to be some reason for being that successful. I started asking the attendees to identify some corporations that were still around and prospering after 150 years, or 100 years, or even 50.

When I returned home and went to the Wellness Center for a workout the two thoughts, of corporate and personal wellness, suddenly merged. People have to die one day, but if they take care of themselves they can keep from rushing into it. Corporations do not have to die, but they do, committing suicide, as a matter of fact, in most cases. The book should be about corporate wellness.

I often receive criticism from reviewers and those who write about quality in particular, stating that my ideas and processes are too simple, that nothing really works out that way. This always amazes me because I only write about what I have actually done. When I talk with people about this, it always turns out that they have not actually done what I have written about. Also they usually have no management experience beyond the department level. They are not looking for a philosophy based on ideas; they are looking for a regime based on techniques and procedures. The ones who respond to me are executives who have had the problem of trying to lead some effort where there were no paths.

Because of both groups I wanted to write about wellness in a way that could not be misunderstood. In a personal wellness center they conduct a profile examination of the individual. They determine the physical status, all the blood chemistry, get a diary of food intake, and some details on lifestyle. This takes a couple of weeks and results in the analyst sitting down with the individual and explaining where that person stands in reality.

"You have high blood pressure, you are overweight, you eat 60 percent of your diet in fat, you smoke, you have high cholesterol, and you do virtually no exercise. The probability is that you will not live to reach 60 years old."

Then the analyst points out that a change in lifestyle could add 20 or more years plus greatly improve the client's enjoyment of life. This change involves learning how to be well and taking some actions to stay that way. Most of these are sensible and none of them are abhorrent. But it is up to the person to determine how healthy she or he is going to be. Corporations have this same problem. They get fat and lazy. Their communication networks become clogged just like human arteries.

In order to make all this clear I built the book around a grid based on how the corporate wellness was in terms of Quality, Change, Growth, Customers, and Employees. To show the current status, I used terms that are familiar in medicine: Comatose, Intensive Care, Progressive Care, Healing, and Wellness. I explained all these in detail and put in stories of people in business trying to help their companies become eternally successful. There was not much about quality management except to show that it was an integral part of general management in any area. The cover of the book said that it was a "new business philosophy," and that is actually what I was trying to put forth.

The book has done well in both hard and soft cover, but it took a few years for the idea of wellness to come across. Even today it is not something most people are comfortable with. It is too much involved with prevention and with what happens today affecting what will be several years from now. I can tell these things from what people say to me when we come face to face. The good news part of being an author is that people seek you out wherever you are in order to share their feelings about your products. Without exception, those who come up to me are courteous and respectful. I don't really care if they agree or not as long as they received some stimulus that made them think about the way they are working.

About this time we received an invitation from Saatchi & Saatchi to merge with them. They were trying to build the largest advertising firm in the world and wanted to add a consulting group to it. They had already picked up a couple of firms and wanted to add PCA to their string. They promised that we would be given hands-free management as long as we met the agreed performance, and that we would be bought out in cash. They were offering about $13 for our stock which was selling at $6 or so at that time. We agreed to talk with them and things were moving along well until the British elections, with which they were deeply involved. This stalled discussions for six weeks or so and then everything sort of petered out. I think they were beginning to realize that they had a great deal on their plate already. They paid all our expenses and we parted friends.

After that another advertising agency asked us to consider a merger. They had figured out that PCA was actually a product-line company rather than a consulting firm. Our courses were taught by people we had taught to teach them, not by those who were expert in themselves. We could teach any qualified person to do that. In advertising they say that the inventory goes down the elevator every night at 5 p.m. With PCA the product was there all the time. I thought it was very astute of them to determine this. We were never able to explain it to the financial community. The deal didn't get anywhere because they were not willing to make a high enough offer. However, as it turned out, we would have done very well if we had accepted their stock in exchange. It has increased dramatically over the years.

For PCA to grow worldwide, it seemed that we were going to need a partner. Setting up in a new town, like Sydney, Australia, for instance, required a big investment and a lot of time to become acquainted. At this time we had offices in Paris, London, Munich, San Jose, Chicago, and Winter Park. Many clients with worldwide operations wanted us to be near them. We daydreamed about finding a partner who was already in places we were not. They could introduce us around and let us use their phone. We decided to just keep plowing along, the company was doing well, although the stock kept dropping. The financial analysts just couldn't figure us out. Their big problem seemed to be an inability to understand our marketing plan. Since we had no salespeople and relied on word of mouth, they had difficulty accepting that it worked, even though it obviously did. My speeches and articles brought people in, but most of the inquiries came from suppliers and friends of our clients. What we did for companies really worked, it really produced results.

I was becoming more interested in writing and speaking than in running the company. Larry McFadin and the other executives did a good job of handling operations, but they were not very creative when it came to new products or relationships with clients. However, they were millions of dollars a year ahead of the competition which was beginning to emerge. The company was safe but it looked like the stock was never going to go up in value again. We discussed buying it all back and even had an informal agreement with a bank to help make it happen. It didn't seem to me like the others had the fire in the belly necessary to make such a buyout happen. Most of them were comfortable already from the sale of stock they already owned. As I noted, we had created a bunch of millionaires.

I received the news from my regular prostate checkup that I was going to need a "roto rooter" job. There were stones in my prostate and the next step would be cancer if everything followed the normal path. So I went into the hospital and had it done. The surgery is not a big problem except that they give a spinal anesthetic. Recovering becomes depressing when you realize that the big senseless lumps you are feeling with your hands are your hips. Then there is the matter of sitting around with a bloody sack hanging from a tube which has been inserted in you. However, it all passes.

The doctor wanted me to be settled down for six weeks. This meant not traveling; I didn't have any problem about not going to work. This offered me a great opportunity, I thought; I was forced to spend that amount of time in my library. I could do some radio interviews by telephone but mostly there was reading and thinking as the main source of activity. Then I received some audiotapes that were made of the open discussions I held each week with the students. These were "free for all" question-and-answer sessions and they had been going on for several years. They seemed to be the material for a slam-dunk book.

I obtained a list of all the questions and cut them down to what I felt were the 96 most interesting ones. Then I proceeded to write answers for each one, without paying much attention to the answers I had given before. As it turned out, many of the replies were somewhat similar, although I had more information now than then. In no case did the philosophy change, but I was often able to offer more practical advice. I didn't tell anyone what I was doing and worked happily along for the assigned time. At the end of four weeks I had all the manuscript completed, on a computer disk. Debbie Eifert, my assistant, ran through the material and checked my spelling and grammar. This was the first time I had ever written a book without any hard copy at all. We then transmitted the content to McGraw-Hill where the editor, Jim Bessent, rearranged the subject matter so the questions bore some relationship to each other. He did this with my agreement, of course. It was a good idea. Since the book was clean, they went right to work on it and in a record time *Let's Talk Quality* appeared on the bookshelves.

It was subtitled *96 Questions You Always Wanted to Ask Phil Crosby*, and Lee Iacocca was gracious enough to permit us to use a quote from his book on the cover. The book took right off and had a nice sale. Readers could find exactly what areas they were interested in seeing. The softcover version is now selling in a dozen languages. It is an ill wind that blows no good.

In 1989 we were approached by the Alexander Proudfoot company, which had just gone from being a partnership to a public company listed on the London exchange. They were American to their roots and were all over the world, including Sydney. They offered us twice what the stock was selling for and swore that PCA would remain independent of corporate authority. They agreed to honor all of our contracts and were particularly insistent that I remain as creative director and chairman for as long as I wished. I let Larry McFadin, the CEO, and our counsel conduct the negotiations except for saying that I thought $60 million was a proper price for the company. I still owned a little less than 10 percent. The PCA revenues for that year were around $75 million, with an after-tax profit of 9 percent. The deal was completed in 90 days.

At this time I was writing *Leading: The Art of Becoming an Executive*. Still trying to reach the unreachable, I put this in the form of a novel with a leading character and a recognizable story. My thought was to show what goes on inside the head of a successful executive. I felt that would help those who were trying to change a company from within. The book also had a new and practical philosophy about the focus of a leader: finance, quality, and relationships. Quality was portrayed as the structure, the body of the organization; Finance is the blood supply; and Relationships the soul.

Those who read it thought it was useful. However, it had such a poor reception that it didn't make it into paperback, the only one of my books to suffer that fate. It did make money for the publisher, however. Later I took the same character and made him the center of a mystery novel. It was rejected with the comment that "We like your business philosophy books because they are clear, readable, and logical. Those are the reasons we don't like the mystery. Stick to business." I keep writing the other stuff anyway.

I began working with an agent for the first time, Al Lowman of Authors and Artists Group in New York. Like all writers I felt that my publisher didn't appreciate me, and like all writers I wanted a larger advance to assure that the publisher would be interested in the promotion of the book. I had been working on the concept that quality management was going to be a real part of management in the twenty-first century. Out of this came *Completeness: Quality for the 21st Century*. It was a little outlandish in some areas, dealing with virtual reality, for instance. It also projected the fall of the U.S.S.R. However, the book has done well and all the things I prophesied are coming true. The idea of *Completeness* was to aim management at making the employees, suppliers, and customers successful. This would let them forget all the TQM stuff which was mainly motivation, activities, and manipulation. In the introduction I made sure the world knew how I felt about all that and the Baldrige Award too. The reaction brings me back to 1962 when most people rejected my thought about doing things right the first time.

I had felt that PCA needed a package for small businesses and began working on it. However, it was becoming apparent that the Proudfoot executives had a different idea of what consulting and education was about than I did. Since it was their company now, I suggested that we arrange for me to retire early. That way everyone could do what they wanted. I would still be available to them for counseling, speeches, and such. They graciously permitted this to happen, gave me a very nice party, and I left. When I left ITT I missed the International golf course at Bolton, Massachusetts; with PCA I missed the people and the airplane. I never missed the business.

My daughter Phylis was still working for Proudfoot as a corporate vice president, but she was interested in working with me in a new venture. We set up a little company called Career IV, Inc., a "Subchapter S" incorporation. Peggy, Phylis, Philip, Jr., and I became the co-owners. Each of us made an investment to start the company. Phylis would be CEO and I vowed never to do another thing from an administrative standpoint. We were not permitted by my noncompete contract to teach Quality Management, but I had no interest in doing so. My attention was laid on the subject of leadership. I felt that the nation was in desperate straits and wanted to change that.

We received many speaking invitations, and although my audiences were interested in quality they seemed to just want to know what I was thinking about. That was an interesting change from the past; they actually were becoming interested in me personally, not just in the nuts and bolts of quality. Phylis put this in perspective, punching my balloon in the process, by informing me that I had become an icon. From iconoclast to icon in a few decades. However, this let me become more comfortable; I could talk about anything I wanted to and it was agreeable to the audience.

Speakers' bureaus provided some of the speeches but most just came from people calling in for organizations. We charged $15,000 and expenses which seemed to be within everyone's budget. We also did some free speeches for groups that were interesting or not for profit. I was able to start each of these with the comment that "Sam (or whoever booked me) said he had heard I was a student of history and asked me what part of the Bill of Rights did I think was the most important?" "Free speech," I would reply. "That's what I want to talk to you about," he would say. This always got a laugh and also let the audience know that I did this for a living, and that this was a special occasion.

I found that many senior management teams were concerned that they were not getting anywhere with the quality programs they were conducting. Everyone was into TQM and a dozen other activity-oriented programs. They would call me and ask that we spend a day together to see if I could suggest what their problem was. I arranged to spend a day with their operations first and the result was that we could have a good "green field" session. In these sessions we look at the company as if it had no people or buildings, just a green field. Then we put on the field what we really need to run the company. Usually it is less than 75 percent of what is there now. I put some of these into the text of *Completeness*. Actually the problem is always the same—management is trying to delegate something only they can do.

Many groups asked me to relate the differences between Dr. Deming, Dr. Juran, and I. I always said that I respected the work of these two men but that we were in different businesses. Dr. Deming, whom I had never met, was a statistician and was the top of his field; Dr. Juran, whom I had met years ago, was a leading expert on Quality Control and quality engineering. Both were worth listening to, even if they apparently made unflattering comments about me and my work. They would say that I wanted to motivate and "exhort" the workers. Since I had never proposed anything like that in any of my writing or speaking, it didn't bother me. The difference between us was that I was dealing with Quality Management, which is not Quality Control or statistics. Also I had learned my trade by fighting my way up through the ranks and actually implementing my concepts in many businesses and cultures. I considered them irrelevant to what I was working on, and I think they returned the thought. We were not very important to each other.

The evolvement of the Quality Management concepts in my mind during my career is something that did not happen to other people. Therefore I am patient, knowing that one day all leaders will understand that it is better to prevent than find and fix. That realization may come long after I am gone. My hope is that it will help produce a world where everyone can have a meaningful job as well as a proper life. People just need the opportunity. The content of my books is built around stories that came from the real-life involvement I experienced. These stories let people relate to the concepts laid out in the first chapters.

THE QUALITY IMPROVEMENT PROCESS REVISITED

Somewhere in the mid-1960s I was asked to jot down the essentials of a quality improvement process. Innocently I ran off the following subject titles which must be addressed. I casually called them "steps":

1. Management Commitment
2. Quality Improvement Team
3. Quality Measurement
4. Cost of Quality Evaluation
5. Quality Awareness
6. Corrective Action
7. Establish an Ad Hoc Committee for Zero Defects Planning
8. Supervisor Training
9. Zero Defects Day
10. Goal Setting
11. Error Cause Removal
12. Recognition
13. Quality Councils
14. Do It Over Again

I really regret the day I dashed off the numbered list of 13. (It became 14 items when I added "Do It Over" in order not to offend the superstitious.) People who write college books about quality always solemnly put the list (14 steps) in a box, doing the same with other people's lists, as if that were the entire content of what we are thinking about. If I had just written them out, with commas in between, and not placed a number next to each of them it might have been better. Many folks looked at the list the way they do a scavenger hunt. They get a commitment from management in the form of a nod or note, and return quickly to place their prize on the contest table. This is followed by a cheerful dashing about, gathering people to be on a "team," mulling over some measurements, deciding to worry about the cost of quality later, buying some trinkets for recognition, and as quickly as possible setting up an entertaining ZD Day. Very little thought is expended as to what the list is about in the first place or the effect it is designed to obtain. Searching for specific procedures among ideas is the sign of the lazy mind, and there are a lot of lazy minds involved in quality improvement. Most of those responsible for setting up a "quality program" were just put in place by someone else. It wasn't their idea.

When people really want something personally they figure a way of making it happen. Everyone who has participated on one side or another of a courtship, for instance, knows that the simple concepts involved in communication between a man and a woman generate a continuous need for original thought and action. Those original thoughts and actions are what make up the bulk of literature, film, and other communication arts. If life management were all cut and dried, Shakespeare would have needed to write but one romantic play; Tracy and Hepburn would have only made one movie. The same is true in political history where the idea is to gain and keep power. The stories of how that comes about, or fails to do so, fill thousands of books and plays. People are quite inventive when their goals are clear and personal.

What happens with quality improvement as a process is that the overall reason for doing it often gets lost somewhere early in the trip. Instead of searching out a future that is bright, prosperous, and free of pain, quality management people get all involved with examining, documenting, and carefully placing the stones that cross the stream. They take pictures of the stones, write stories about them, hold conferences on their merits, but hardly ever get around to actually crossing the stream. Then of course there is the complicated process of selecting and packing proper clothes, studying dance steps that are likely to be involved, all the while arranging the purchase of necessary tickets. The result of this muddled effort becomes small pockets of progress which are happily reported to their eager management. This is like being thrilled that our eight-year-old made it home from school without falling into the pond. Who cares whether she learned anything, built social skills, ate a proper lunch, or grew a tenth of an inch?

"The more detailed information you give people, the less they accomplish." That is my quote; I wouldn't want anyone else to take the blame for revealing this truth.

The idea of having a formal process, in anything, is to take advantage of the thinking that has gone on before. When we go to school it is not necessary for us to invent arithmetic, or sentence structure, or even the fox trot. All of these subjects have been done, documented, and as such can be taught and learned. We can begin with that knowledge already at our disposal. We leave school with these basics as part of our weaponry to make our way in the world. We add to this knowledge through experience, study, and reflection. We keep learning, whether we deliberately set out to do so or not. Some things we think are difficult, some we think are easy. Here is the way I feel these subjects should be viewed, and handled.

MANAGEMENT COMMITMENT

Quality is something that always starts out to be easy. Someone decides that something should be done about it and summons someone. This is the Marshall Wyatt Earp complex. The town is full of rowdies; decent people aren't safe on the streets; something must be done. Call Earp. He will bring order from chaos, cause vesper meetings to be held on Wednesday evenings, turn rowdies into productive, serious-minded citizens.

How is this brought about? Inflexible rules and the force of arms. "No guns are permitted south of 4th Street"; "Liquor will not be served to drunks"; "Those who do not pay their just bills will be jailed." Town meetings are held and the "decent" people are forced to take a stand against the evil ones. In the movies it is not hard to separate the two groups; real life does not necessarily paint things that clearly. Often the one who is the biggest problem appears to be the biggest helper. (I have found this to be true in many real-life cases.)

At this point the management, those citizens who hired Earp, begin to find that "management commitment" really means the chance of getting one's hands dirty. They have to get up on the rooftops with their rifles to help the marshall face down those who prefer the life of disruption and zero rules. If they do not do this, then the bad guys, and the good ones, will know for certain that the management commitment was shallow and tentative. The store and bar owners who profited from said disruption often find it hard to make the choice of mounting the stairs, rifle in hand. They would like to mediate and arrange to permit just a little dastardly nonconformance, like when you have the proper amount of cream and sugar in your coffee. It still tastes like coffee, but it doesn't taste too much like coffee.

All of this brings out another truth: "Not everyone really wants to get everything done right the first time." Now, can anyone possibly be against quality? one might ask. But it might be well to wonder, as a corollary, why quality is a problem at all. If it makes so much sense to get the right thing done right, then how come the normal way of operating almost everything is to do the wrong thing first and the right thing second? Why does every town want to find a marshall to get the citizens to do what they already know is best for them? In the movies he usually has to shoot a dozen or so of the bad guys in order to bring all that about; after he leaves, it begins to creep back in. Usually the next coming is not as bad, since some of those who caused the problems are no longer with us. But there it is, growing back out of the ground, needing to be crushed underfoot once again. Success always brings about the conditions that breed failure. Hardly anyone repeats as conference or league champion. Arrogance is the child of overconfidence.

For some reason many executives think that when they decide it is time to call in the marshall, their management commitment is all over. Once they find themselves standing on the roof with a real, and loaded, rifle waiting for the real, and genuinely hostile, bad guys to ride in, the true meaning of management commitment begins to reach them. If they do not move to this level of activity, everyone from bad guys to good guys will know they are just fooling around. No one is going to put their neck on the line for someone who is just fooling around.

QUALITY IMPROVEMENT TEAM

The business of having a "team" to direct the concert of improvement has all really become confused. Just having a team doesn't mean the players can play whatever game comes up. It was never my intention that the team should run an improvement program; I was just talking about communication. Earp is the one who actually presses trigger against flesh and shoots those who refuse to obey the laws. Hiring an Earp is only effective in fiction; real life requires that the very senior level of management strap on the weapons and lead the battle.

In this regard my calling for a team had two purposes: the first was to force the thought leaders to come together and face the fact that quality was a problem and that they were the source of it; the second was to provide a conduit for communication between all those departments who never paid much attention to each other. I always insisted that the group gather formally and direct the actions that were to take place in order to cause improvement. Usually they would appoint some young talented person to organize them and the effort into a program. This person then, being no dummy, realizing the opportunity presented to him or her, immediately turns the effort away from the "team" and concentrates on the people. This action dooms the process unless that person is smart enough to recognize management needs to be included in everything or the group will go back to sleep. This objective is reached by having "quality" placed as the first item on the monthly management meeting. This forces each of the managers to become involved enough to talk about what is happening or not happening. The CEO then needs to focus on the company's business process, beginning with interactions with the customer. What are our problems? What do they cost us? What are we going to do about them? This dialogue serves to force action that otherwise would not take place. It makes the management team address the reality of the need to actively manage quality, not just let whatever happens happen.

QUALITY MEASUREMENT

Faced with the need to understand exactly why quality is a problem, and exactly what that problem is, the team members begin to concern themselves with measurement. When they deal with the real world they begin tracking customer complaints at the back of the company and work their way forward, slaying dragons as they go. They should begin with the negative things: What are the customers unhappy about? Delivery? Conformance? Attitudes? Price? Let's get very specific about these problems so we can get specific about the causes. What are these causes? Training? Wrong people? Inadequate requirements? Poor management direction? Customer misunderstanding? Supplier problem? We can then attack these causes and begin their elimination.

A permanent measurement system comes from setting up a Complete Transaction Rating as described in Chapter 5.

COST OF QUALITY EVALUATION

I quit calling this "Cost of Quality" some years ago because the quality professionals reacted negatively to it. They conceived it as being blamed on them and thus resisted doing it at all. When forced to have it documented, they would only list enough things to report a low number, like 3 percent of sales. This, of course, scared nobody. But when we call it the Price of Nonconformance we have now defused it. Every transaction in the company is accomplished either in a conforming or nonconforming manner. Black or white, nonconforming means that it has to be done over, or corrected, or something. That costs money that was not planned to be expended. It has to come from somewhere, and that is usually reduced margin, which means reduced gross profit. In manufacturing the price is typically 25 percent or more of revenue; service companies spend 40 percent or more of operating expenses. When these numbers are shown—by department, unit, division, region, nation, or any other business community—everyone pays attention. Even the most reluctant of executives will get moving when faced with this reality.

QUALITY AWARENESS

Everyone has to know what is happening. Many companies worry about their unions, but there is no need for concern if they are brought into quality awareness up front. I always insisted that the senior executives of the company and those of the union attend class together. This was a good opportunity to privately remind the company officers that their union counterparts had been elected while they had been appointed. The purpose of quality awareness is to let everyone feel that they, as individuals, belong to this new company attitude. Giving them information in a positive and open manner sets the tone to change the culture to one of care and prevention. Many managements are concerned about telling their suppliers, customers, and employees that they are going to take action to improve quality. Actually all of them, like the union executives, will be delighted. It is not necessary to state that "we have been terrible and are going to get better." We can talk about the desire to give customers exactly what they want; we can talk about the need to be competitive in this world economy; we can talk about individual contribution in terms of ideas and actions.

Newsletters, Family Council meetings, rallies, open conversations, articles, news stories, all the tools of public relations need to be applied. The purpose is to communicate, to make aware that change is in progress. There is no need to attempt to convince through awareness. That comes from education and experience.

CORRECTIVE ACTION

The need for organized positive action will be seen immediately. As soon as someone determines, for instance, that customers are complaining because the bills they receive are incorrect and not complete, corrective action is required. Since the goal is to have all transactions correct and complete, we must do more than go yell at the Accounts Payable people. We have to examine the process used to calculate, prepare, and send bills. Why, specifically, is it inadequate? What specific management action is necessary in order to correct the present situation and prevent problems in the future? These questions raise the concern for answers that will resolve the situation.

The idea of a formal corrective action system is to get everyone in the habit of solving and preventing rather than learning to live with problems. It is easy to determine if a company is used to not taking care of difficulties: they know how to get around all the problems. They have been there before.

Establish an ad hoc committee for Zero Defects planning. We want to bury the idea of complete transactions deep in the culture of our organization. To do this we have to show that management is serious about it all, serious enough to donate their bodies to the cause. There is nothing motivational about Zero Defects; it is just plain old specific communications. My father used it on me many times on such subjects as taking the ashes out of the furnace, and then basement, without spilling any on the floor. He didn't know to say "zero ashes on the floor," but I got the message. If he had not done that, I would have not known the standard. When people come to work they do not automatically know that doing things right is the way things are done around here.

This committee needs to arrange a communication between those who establish the performance standard and those who perform it. The key part of this arrangement is to make it clear that management has internalized Zero Defects and will be acting in that way themselves. They will begin, and end, meetings on time; they will be disappointed when reports are not delivered as promised; they will ache when a customer is treated improperly; they will congratulate those who have done right and encourage those who are trying to get it right.

This is a serious activity. Some implementers think that getting the employees to sign up for ZD is what this is about. They have not given the subject much thought. The real purpose of ZD planning is to place management in a position they cannot wiggle out of. If they are permitted to change in and out according to the stresses of the day, ZD will be a myth.

TRAINING

Originally I thought that training the supervisors (which includes the chairperson) to understand the concepts of quality management and their personal role in the process was enough. However, it soon became apparent that this was far from enough. Supervision, at all levels, from the chairperson down, seems to be brain dead when it comes to passing along performance standards or job understanding. We need to give each person the opportunity to understand what quality is all about, and what their role is in this equation. They need to internalize concepts; they need job skills; they need access to information. They don't need activities or lectures. Most of the quality education classes are about things that make management think something good is happening. People need serious information that relates to the way they make their living.

So education and training must be accomplished in a planned fashion and be conducted relentlessly. It cannot be a once-around-the-circuit program; it has to be continual. As people progress in understanding and implementing prevention, they need to know more about it. When I bought an organ in 1961 I took the six lessons that came along with it and got to where I could read the music and play rhythm. That was the end of my education in organ playing. Today I play exactly the same as I did in 1961. It is satisfactory for me, and does not irritate anyone who wanders by, but it is not very good. And my organ playing will never get any better until I learn more about it, and practice what I have learned. So educating people about concepts while supplying updated information, and training them to accomplish tasks, must be a long-range effort. The more actual working folks can be used to conduct classes, the better off their students will be. Nothing succeeds like real life.

ZERO DEFECTS DAY

The idea of having a celebration day was born to implant the message that things were different around here. I always suggested that the quality improvement team arrange one event every year. Usually they only take an hour or so away from the job, but the reminder goes on all day. The office is decorated a little, a children's poster contest may have been held, a special cake is around, a customer is asked to come speak, everyone signs a poster of rededication, and management walks around smiling. There are just enough interesting things to show that this is a different day and help people feel happy about their work. It is also a good time to present awards, peer nominated of course.

Some people make fun of having a celebration day because they think that it is the whole ball of wax. Those same folks go happily to weddings, christenings, recognition dinners, inaugurations, and such. It is the same idea, and it is very effective. However it must be a part of the whole; it is not a whole all by itself. Otherwise everyone could go back to their own house after the wedding and not worry about the future.

GOAL SETTING

People often consider the practice of goal setting to be some sort of ceremony, but it is a normal part of everyday life and conversation.

"I want to be at the hairdressers by five."

"I am going to balance my bank account tonight."

"Next week I am going to call seven customers."

"I am going to lose some weight."

The first three of these goals have a chance to happen; the fourth is not specific. Goals have to have numbers or specific objectives in them. "I am going to lose a pound a week until 15 of them have gone" is specific. One can lay out a plan, a process of operations. When an organization decides to encourage its people to set goals, they only need to teach them this lesson about specific, and measurable, objectives. It is not necessary to set up some complex system.

The best way to work on goals is to incorporate them into personnel reviews. When the employee and supervisor sit down to discuss performance they need something to talk about that is not subjective. Relationships between people are based on all kinds of things that they may not even know exist. Personal chemistry is something that cannot be predicted or measured.

Unfortunately, that is the basis of most reviews. If we are unlucky, we can do good work and get bad reviews. This builds a reputation for us which clouds all future work. After all, it is written that a person who develops a reputation for being an early riser can then sleep until noon.

For this reason, evaluation should be based on the accomplishment of agreed, and clearly measurable, goals. It is up to the leader to make certain that the goals are worthy, and it is up to the employees to make certain that they want to do them. These goal discussions then help define everyone's understanding of the job assignment and fit in nicely with the Complete Transaction Rating concept. The advantage to the organization is that they can now know who really does good work and the employee can know what good work is.

ERROR CAUSE REMOVAL

The idea of error cause removal (ECR) is to help individual employees cause corrective action that affects their work. We are asking everyone to create a Zero Defects environment around the workplace, but often there are blocks to keep the objective from happening. People need a way of eliminating these blocks. My experience as a line worker was that no one took my complaints seriously. For that reason I created ECR when my chance arrived.

The way ECR works is that a simple, one-page form is created and placed where everyone can obtain it. (The supervisor's desk is not the place I have in mind.) As part of the regular quality classes it is explained that the purpose of ECR is to let people state a problem. It is not necessary to offer a way of solving it, but a solution would be welcome, of course. Then the form is put back in the box where it is picked up by the quality improvement team member and placed for action. The person who filled out the form is told right away that it has been received. As soon as some action is planned, that person will be informed of it; he or she knows more about the job than anyone and needs to be encouraged to trust the ECR system. Often the person responsible for corrective action on the item will come back to the writer for additional information. Probably 75 percent of ECRs can be corrected at the supervisor level.

It is essential to remember that the ECR system is for communication and that the individual must be included in the cycle. There will be a lot of these types of programs if the people involved are treated with respect. Suggestion programs have a very low response—probably less than 4 percent of the people participate. Because of this fact, there is always a fuss about awards to suggestion givers, particularly if those awards are financial. ECRs, when treated with dignity, will involve 100 percent of the people in the organization. Then it is possible to have little rewards, like drawing a suggestion at random so the maker of that suggestion can use the number-one parking spot for a week. I recommend using ECRs actively for a month or six weeks twice a year. Then twice a year use Make Certain which has a different twist. That twist is discussed in Chapter 14.

RECOGNITION

Everyone wants to be appreciated, but they want that appreciation to be genuine. This means it must be dignified and it must come from peers.

One of my children went to a camp one summer. When it came time to return home we went up for the final session and they presented awards. There were 80 or so kids in that camp and everyone got at least one award, all selected by the counselors and management. After a while the presentations deteriorated into a droning from the platform and mumbling from the audience. The last award was for "most popular camper" who was selected by the other campers. When the young lady's name was called, all the other campers cheered and rose to give her an ovation. Before we left I chatted with the camp director about the difference in audience reaction, but he had no comprehension of it at all. He did say that the same thing happened every year.

Awards should be meaningful. They shouldn't be some form of money; they don't have to be valuable. Military medals are trinkets but well respected because of the way they are presented and treated. People do not nominate themselves for such things. In ITT I started the Ring of Quality program which is still going to this day. It is explained in detail in *Quality Is Free* so I won't go over it again. But the important point is that in these 20 or so years no one has ever questioned a selection. All of them were nominated by peers, all selected by peers. This is my primary problem with government quality awards such as the Baldrige. If you have to nominate yourself, and even pay for an evaluation, what worth can it be?

The proper way to determine the recognition in an organization is to bring together a temporary group to decide what type of recognition should be done. This group should represent all levels of the organization and should not have a senior executive who will dominate the proceedings. If I have learned one thing in all these years of sweat and toil (well, perhaps not too much sweat), it is that management always picks the wrong people to recognize with the wrong stuff.

QUALITY COUNCILS

This action was set up for companies who had quality departments in several areas. The idea was to bring professionals together in order to have them working toward common, and agreed, goals in a proper manner. In ITT I used the councils as a volunteer organization which did the management job in which an official system would have failed. This is explained in *Quality Is Free*. I learned, once again, that people work enthusiastically when they feel they are a vital part of the operation, and they are slothful when work is imposed on them.

Council memberships do not have to consist of professionals; all that is needed is a common interest in some subject. Quality Councils are different from "teams" in that they are essentially making functional policy while accomplishing the educational and self-help activities necessary to support that policy. So instead of sitting around and complaining, they get to make things happen. The wise leader helps arrange their agenda in a quiet fashion.

DO IT OVER AGAIN

Actually the quality process goes on forever, if it is to do any good. It is like eating properly. Once our ideal weight is attained, the battle really begins. Selecting a lifestyle that produces the desired results requires eternal vigilance and action. Commitment must be reconfirmed every day; that is why happy couples wake in the morning to smile at each other and state: "I love you." It has to be done regularly.

Our budget cannot be permitted to exist on its own without attention for even a moment. One wild bit of credit card forgetfulness will place us in a deep black hole that requires months to egress.

Driving to Grandma's house requires making a certain amount of travel and turning. If one movement is omitted, it means that a great deal of effort and gasoline will be expended for inadequate results.

When I suggest doing it over again, I really mean the whole business has to be repeated. A new Quality Improvement Team should be appointed and the baton passed from old to new. The whole team should go; leave no one on for continuity. The old will think there is nothing left to be done, but the new will find dozens of actions left unfulfilled. I like to have a transition ceremony at a Family Council meeting just to make certain everyone knows what is happening.

CHAPTER

THIRTEEN

THE TOUGH COMPANY

Recently I had lunch with the CEO of a large company in what he called his private dining room. Actually it was a small conference room with a table for four, file cabinets in the corner, and a view of the backyard. Lunch was a tuna salad ordered in from the neighborhood deli. This is a far cry from the way CEOs used to live back when I was working in New York. I have had many meals with crystal, silver, and real china, served by muted bodies who respectfully whisked dishes in and out. In those days no respectable company was without a huge mahogany table and sideboards where a variety of good food and drink were displayed.

Here we sat in our shirt-sleeves and poured our own drinks, mine a bottled water, his a diet Coke. The delivery girl from the deli brought the salads into the room and the CEO paid her from his pocket. I halfheartedly offered to remit my share of the lunch but he waved me off saying that he would stick me with a real dinner check one day. We munched away chatting about current events and remembering a time when we had both worked on the same project for different companies.

"You get to see a lot of companies, and talk with many people," he said. "Who would you recommend that we go study? What companies do you consider to be wonderful?"

It is difficult to keep from laughing when someone asks that question, and almost everyone does. Companies, like people, are continually changing feet of clay. At this given moment they might be "wonderful," but who knows what is going to happen in the next moment? By the time I would finish describing the company I thought was doing everything properly, the news would arrive that their board had been indicted for fraud.

I read *The New York Times*, *The Wall Street Journal*, *The London Financial Times*, and the local paper every day. I read all the business magazines, I listen to business news, and I talk to people all around the world. My analysis is that there is no "wonderful" company. There are many that have been successful for a while but few who have maintained that status. Everyone is either coming or going, and sometimes they do not know which track they are riding.

We must be careful when we talk about "success." Being profitable is not enough; having an expanding market share is not enough; none of the traditional measurements we hear about on the seven o'clock business news are enough. Success requires that all the necessary characteristics happen at the same time, in parallel. And it has to be pulled off over a period of time. The organization must thrive through recession as well as boom. It has to be able to deal with whatever shows up. It has to have a clear agenda, an understandable philosophy, and a worldly posture. All of this comes from leadership. I call this kind of an organization the "tough company." It comes out on top every time because it is resilient.

The CEO poked at his salad. I was getting the impression that this dissertation wasn't much help to him. He was looking for a package to insert, like a computer software disk, into his company.

"'Tough company,'" he repeated. "That sounds like another one of those slogans that comes along regularly."

I allowed as how (as we say in West Virginia) it could be construed that way but also stated that it was a description that could be understood and didn't relate to some sort of system. All I wanted to do was describe what had to be accomplished. How that was done required thought and innovation for the unknown, along with pragmatic application for the known. I have owned a dozen cars, for instance, and had to learn their eccentricities. But my driving skills were applicable to all of them.

"I'm not certain I understand that analogy," he stated. "I think you are being gentle with me, and I need more specifics. How about I ask my key people to join us and you tell us the plain unvarnished truth, as my grandmother used to say?"

I asked if he wanted to know what a tough company was and how they could get on that track. He did. Two men and two women joined us about 10 minutes after we had poured ourselves some coffee from the thermos. The CEO introduced them. All four were obviously bright, which is one of the criteria for getting to that level. Two were in operations, one ran finance, and one did human resources. All of them had been with the company at least four years. We did introductions, discussed things we had in common, then they all sat back and stared at the CEO.

"I asked him about successful companies and he started talking about 'the tough company'," he said, nodding at me. "Apparently that is what a successful company is. So I asked you all to come sit with us while he explains what a tough company is and what we have to do to join the ranks."

With that he smiled and they all switched their gaze to stare at me. I began by pointing out that we were going to look at things in a different way. We should not wrap ourselves up in things that were well known, or that had survived the "test of time." We should remember that at one time Trinity College of Cambridge University made it an offense punishable by expulsion to study anybody's work but Aristotle. It turned out that there were a few things in which the great philosopher was in error. Many things, such as the sun and planets revolving around the earth in perfect circles, were found to not be true. He just did not have the equipment to verify his thoughts. When the seventeenth-century enlightenment came along it was because people began to use experiments and reality to determine what they knew and did not know. We are going to talk about the broad aspects of leadership and management. We need to do it with an open mind.

"I think you will find all of us have open minds," smiled Human Resources. "That is practically a policy around here."

I shook my head and told them that their quality director had shown me the corporate quality manual which was strictly lined up with the criteria from an award program. He indicated that he had a hard time getting anyone to listen to anything that wasn't on that list. That represented a closed-mind attitude to me, but perhaps they had another definition.

"That quality criteria was put together by experts," said Operations. "Why should we question it?"

"Just like Cambridge?" another asked.

I suggested that they had no way of knowing whether these people were experts or not. Criteria like this makes a good checklist, perhaps, but limits thinking. I asked if they could think of a single thing, in any field, that could happen all the time, everywhere, with one criteria. I told them that I had an income tax expert who was considered the primary authority in his firm. Last year I corrected him on something. I knew about this change in the tax law because a golfing friend had told me about it. My tax expert was reluctant to look into it but finally did and found it all to be true. His mind wasn't open either. If you lock yourself into some guru on a subject, you will get bypassed eventually. Also, a lot of people who write and teach on something like quality have never done it for a living. If I am going to walk across the Andes, I would like someone with me who had actually done it before, not someone who just thought about it. Theory won't cut it when we have to face reality.

"But a lot of things have been proven in the past, and have actually been considered to have stood the test of time," said the CEO. "Should we toss them out because they are old?"

Hanging around doesn't necessarily mean something is correct or useful. People have often had the wrong beliefs for thousands of years, and have defended them with their swords. Being old is no guarantee that they work, or ever worked. It also doesn't necessarily mean they are wrong. We have to look at everything in today's terms and decide how realistic and useful it is right now. Age can be an advantage, but not for everything. Most companies are managed according to conventional wisdom, which includes the old standards and many of the passing fads. But how many companies do well? How many current leading businesses were in that position a few years ago, and how many will be there in a few years? There is no one way to manage; we have to learn how to deal with the river as it flows along, not as it looks in a painting. We have to accomplish certain actions, and we have to have an understandable philosophy of operating and communicating. We have to deal with reality.

There was a man who complained of being lethargic, and having no sex life. His doctor told him to walk seven miles each day and then call him in a week. When the man called he reported that he felt great, better than he could remember. The doctor asked about his sex life and the man responded that since he was 49 miles from home he had no way of knowing.

If we are going to exercise by walking or running we have to have a philosophy of doing it that brings us back to our starting point when the trip is over. We all have to understand that the same way. When someone in our relationship is going out, we have to be able to know that they are coming back. If we can do that as a management team, then we can work together without holding hands all day. The really great generals learned to manage in this manner. They were so well oriented to their subordinates that one could know what the other would and would not do.

"Tell us what a 'tough company' management does," asked Human Resources. "Then maybe we can understand what all these little stories mean. So far what you are telling us is what we already know."

I said that must mean that you are operating in that manner then. The little bit I have been around here did not reveal to me that all of you share, or understand, a common philosophy or have agreed upon an agenda. You seem to just do pretty much what happens and then rationalize it later. That is what money managers do; they explain what happens on the stock market after it happens but are unable to predict or control much.

This was greeted with silence and shuffling feet. I really didn't know for sure that they worked this way, but since that is how most companies perform I thought it was worth the effort to say it. Apparently my thought was at or near the truth since no defense was offered.

I don't know exactly how to create a tough company, I told them, but I can describe one; it seems to me that if we can paint its picture we would be able to figure how to create it.

"When you talk about a tough company are you mostly considering manufacturing?" asked Finance. "We are a service organization. It may be different for us."

It would be necessary for us to agree on just what a manufacturing company is and does. Suppose I describe an organization that procures dozens of components; processes the components through heat, cold, and other change mechanisms; combines the components according to specifications; and when it is all over presents the finished product in a box to the customer. Is that a manufacturing company? They all agreed that it was. I asked if McDonald's was a service company, and they agreed that it was. I suggested that I had just described what McDonald's does. Their star product, the Big Mac, consists of several components, each of which has been processed before the assembly is complete. That is manufacturing. They also do service. Most companies are like that; it is not really possible or necessary to separate the two. The question of how many jobs are manufacturing versus service is academic at best. The days of white and blue collar are gone. All we need to deal with is what it takes to run our company.

"So what does it take?"

Companies fail because management becomes arrogant, usually from a period of success, and forgets to listen to employees, suppliers, and customers; they fail because the resources of the company are squandered on things that are not necessary for the basic job; they fail because they do not innovate; and they fail because they are not consistent.

"Those don't sound like earth-shaking items. We could all probably come up with the same list," said Operations.

That is certainly true, but that does not make the list any less pragmatic. People get heart disease for a very short list of reasons. The leading cause of lung cancer is smoking, just one simple item. Overweight and stress bring high blood pressure about. If you spend more than you make, you will have financial problems. The reasons are simple but overcoming them takes policy, education, and hard work.

Let me give you some of the portrait of a tough company. It would be nice if it would work out to use the letters in "tough" to make it all easier to remember. But you all should be able to handle five things. I have seen some of the complex financial data you normally relate to, so my little items should not be too difficult. The portrait of a tough company:

- Finance is focused on results.
- Management knows the customer.
- Employees are veterans with a high CTR.
- Distractions are avoided.
- Management understands a common agenda.

First of all, let's take finance, which I consider to be the nourishment and behavior center of the organization. The key item is what we buy with our money. Everyone knows it is necessary to have revenue of two dollars in order to have a dollar we can do something with. We can also eliminate the need for the two dollars if we don't spend a dollar. Enough money dribbles out of the normal organization to double their profits without doing anything else. Money is spent for things that are not absolutely essential to the agenda of the organization. How often do we see a company forced to lay off 20 percent of its employees and find that it can continue to function, perhaps better? Why did they need them in the first place, if they can now get along without them? We should never have to let anyone go for financial reasons.

During the 1980s I was regularly asked to speak at strategy sessions being held by large companies. They all had Strategic Planning Departments and produced multicolored brochures explaining where the company had been and where it was going.

They would meet in nice resorts, with recreation available; I always enjoyed those sessions. There were well-known speakers whom I found interesting to meet. The entire conversation sounded so solid and useful, it seemed that those corporations would go on forever. However, as far as I know all those departments are gone. Strategic planning is best left to those who have to live, and die, by the results. It is like quality in the old days, when the general manager could just fire the quality manager when customer complaints became too loud. That gave the general manager a chance to go do what should have been done in the first place, and what the departed QM had been advocating all along. Talking about facing bayonets is not the same thing as reality.

Those who build comfortable offices and buildings will find that it is difficult to get people to leave them in order to go where the action is taking place. Management must take pride in creating a "no frills" office and travel environment. While having the latest in office equipment such as computers, multiuse copiers, telephone systems, and satellite communications, they should have work-oriented offices. People notice all this. Waste is contagious. When everyone sees management saving paper clips, they begin to get the message. One of the great tools for preventing waste is the Price of Nonconformance (PONC). Companies that calculate and announce PONC set in motion a great deal of preventative action. This is not just platitude; it really has an effect. We all learn, usually the hard way, what it costs in time and money to run out of gas in some remote location and have to call for help. How much better it is to refill the tank at the correct time. This complete transaction pays off, as my parents told me in advance.

In order to put this way of thinking into a portable package, I would recommend that we use the analogy of a race car. The car itself is the end result of all the effort that the racing company expends. In that same regard the organization we are managing is the end result of our efforts. Money spent on items or actions that do not lead to the car running faster and safer are just plain waste. Worse than that, they make no contribution. Focus on the result.

Now, just to toss another situation into this neat little thought of being cost conscious, let's suppose that a fantastic brain or salesperson comes along. The brain knows how to alter the airflow over the car with the result that the vehicle will reduce fuel consumption and increase its speed at the same time. This is a result that you really desire. However, the brain likes to live well, preferring to fly first class, have hotel suites, and be driven to work in a limo. So does the salesperson who is the top performer and wants a Mercedes SL as a company car. With these situations you have to make a cost-effect trade-off. This means being able to determine the relationship between the individual's personal contribution and the result for the company.

Second, let's consider management knowing the customer. As we can all recognize, those companies that have lost their way usually do it by becoming separated from their customer. Management, particularly, in larger organizations gets so involved with running the place that they turn customers over to the professionals. One would think having a dedicated Marketing Department, an energetic sales organization, an integrity-filled quality process, with continual formal auditing of all, would produce sensitivity to the customer. One would be mistaken. People in such organizations are all involved with their own agenda, and the more successful the company has been in the past, the less they will feel needs to be implemented for the future. If there is doubt about this, just listen to these folks speak of the customer in the abstract. It is as if those who fork over real money for what we provide are invisible. They will use the customer in conversation with management only as a leverage to get something they want. They tend to think of the customer as being one large homogeneous group, all acting and dressing alike.

"The customer prefers to use our charge card."

"The customer is under six feet tall."

"The customer is turning away from fast food."

Managers who want to keep their jobs and their track record

have to literally get out and rub elbows or tummies or something with those who actually purchase their product or service. They cannot obtain this secondhand. Asking the regional sales manager to provide an honest evaluation of why revenues have been dropping over the past two years will produce only erroneous excuses. Competition will be blamed for its unscrupulous ways; customers will be laying back due to economic reasons; it is the wrong season for all this. Nowhere will the real reasons appear, such as: our distribution system is too slow; our credit policies are too tight; we are not reaching out to the customer, just waiting for them to order; the salespeople are only making six calls a day instead of eight; our products are getting out of date; the advertising is stodgy and uninteresting.

They don't know these things because they do not talk to the customer about that kind of stuff. I am not certain it is possible to get them to change their ways along this line. Management needs to conduct their own surveys, using whatever means are available. They are the only ones interested in the customer; they are the only ones interested in the whole company. Everyone else, and I mean everyone, only cares about their own area and its interfaces. That is why children can hardly wait until it is time for them to go set up their own household, be it ever so humble.

Third, is the matter of employees who do the right things right the first time. Those who will accomplish that objective are veterans, with proper training and leadership. In order to acquire veterans who will have high Complete Transaction Ratings, it is necessary to take certain steps. These are selection, training, encouragement, and positive relationships. The problem with selection is that most management does not recognize talent when they see it and most Human Resources organizations would not be able to pick Einstein out of a lineup. They are just not aimed that way so we wind up with neat records of high employee turnover. We must get Human Resources interested in selecting good people and retaining them. You will never find this on a convention agenda.

Training is not a one-shot item. The content of work changes continually; office technology is always on the rise. If management wants to see how well their people are trained, one quick test is to see how many can set up a conference call on their telephone system without asking someone. Training must be continual, it must be formal, and it must be conducted by people who really know what they are talking about.

Encouragement comes from creating an environment where people can enjoy their work life and can have a feeling of accomplishment. No one has ever been able to create a room clean enough to satisfy their mother. I have always felt the natural reason for this situation was to create an incentive for them to leave home one day. People work for satisfaction, and their leadership has to lead them to making this happen. Award programs are nice but personal appreciation is the best. This doesn't mean management has to lower its standards or become glad-handers, but it does require noticing what is going on. Encouragement only requires a pat on the back or a gesture now and then. But it has to be based on real results and be conducted consistently. Managers and supervisors have to be trained how to know about this. I had a boss once who made a practice of taking his staff down the street for an ice cream sundae at random times, perhaps three or four times a year. They learned that this happened only when they were working well together.

Positive relationships come from management's desire to help employees be successful in their work and lives. It is necessary to recognize that employees are in a defensive posture throughout their work lives. Hardly anything about their employment is under their control. Some person they never knew existed, sitting in a small office in a large New York building, can dispose of them with the mark of a ballpoint pen. Whole office complexes can be wiped out when the business strategy changes even slightly; chains of stores drop off the 300 smallest ones over the weekend; pension plans are looted by mergers; and of course we can get old.

All of these situations are the reality of business and must be faced. But if the employees feel that the management is taking their well-being into consideration, is helping them be prepared to do a good job, and has a sensible benefits program, then the beginning of a relationship exists. There are programs, of course, to help all this along. But the role of senior management is more vital here than in any other place in the organization. All employees are extremely sensitive to the way management feels about people. If management thinks some minority group members are not good workers, or if they feel women should not be in top management, or if they think certain types of customers are ignorant, or if they have any other bigotry, it will find its way into the way the employees operate. This is becoming clear in cases of sexual harassment. Often the man involved (and it usually works that way) will be surprised that what he has been doing is considered to be wrong. The effect usually is to create a "blame the victim" attitude. In this type of environment it is difficult to build relationships regardless of how many policies, practices, and nice benefits exist. People know whom to trust.

Fourth, is the matter of distractions and the need to avoid them. If we get all involved with placing sponsors' labels on our race car, we can forget the real purpose of the machine. The entire staff will be drawing views of the car and moving labels around while rubber seals are deteriorating through lack of proper lubrication. Sponsors will want to visit the pit areas and get to meet the drivers. The pit crews can become involved with comparing different products used in maintainance of the automobile. All in all, the distraction will reduce the effectiveness of the whole team. This is why teams making their first trip to a bowl game have trouble concentrating on their play.

A well-known professional golfer I see now and then is coming off two years of results that were not consistent with his previous career standard. When asked about this in the casual atmosphere of a locker room he explained the problem in a way the press had never heard.

"We have been building this house," he said, "and the details kept changing. It was impossible for me to concentrate on my game like I had done in the past. But the house is finished and I am back on track.

"Writers have been talking about my back problems, my vision correction, and a bunch of other things. But all of those are just fine; I can strike the ball as well as I ever did. But when your mind wanders, nothing works right in this business. I have actually only been off about 1.5 strokes per round. That difference is enough to turn luxury into starvation. The thought of being hungry again got my focus back."

I have watched corporations slide down after building a great headquarters building. When CEOs get to be famous or rich they become distracted and drift away. A company jet can make that happen; it doesn't take long before someone begins to think that jetting about is their justified right. It is up to a management team to keep everyone humble enough to avoid confusing their agenda. Of course, the ultimate distraction is to conduct an acquisition or merger. This move hardly ever works out because the purchasing executives never really understand the company they bought. They don't know why it works. They think they do, but they don't. If they thought they didn't understand, and wanted to learn why the company worked, then it probably would come out alright. There is no arrogance like that shown by the executives who buy another business with the idea that they can run it better.

Fifth, management understands a common agenda and philosophy. During my ITT years I reported to what was called the Office of the Chairman. It consisted of three people in one organizational box: the chairman (Harold Geneen), the president (Tim Dunleavy), and one executive vice president (Rich Bennett). The idea was that you could talk to any one of them, agree on something, and the other two would automatically be in agreement. Making this happen required that these three have common goals, agenda, and philosophy. They had to work hard at it in order to make this viable. They were able to pull it off because they thought alike and they talked about agendas, goals, and philosophy. In quality, for instance, they were locked into people doing what they said they would do. Deviation and variation were not considered to be a normal part of life. This made everyone's job easier. Instead of spending time and money getting around agreed requirements, the operating people concentrated on getting the job done properly.

In order to have common understanding, the management team must create policies, written or agreed, that bind them together. It must agree on things like approval for expenditure of money, hiring of employees, standard of performance, typical trip expenses, and many other things, some important, some seemingly trivial. But, as in a marriage, there are no trivialities; everything is up for agreement. The CEO needs to lay out specific policies on quality, relationships, and finance as they apply to employees, suppliers, and customers.

"So we can deliberately create the management environment that produces a 'tough company'?" asked the CEO.

"And no one really is involved except the executive committee?" asked Human Resources.

All true, all true. There is no need for a program or announcement or anything. But the first part of being a tough company is to recognize that great management with a product or service no one wants gets nowhere. If you have a wonderful product or service, then the management doesn't have to be so wonderful. Take a good look at what you are selling, how that is being done, and how successful your customers are. Management teams who are in love with their product create problems that cannot be solved. A tough company is one that can manage anything. Make certain that your "anything" is worth managing.

MAKE CERTAIN

I wanted to retain the chapter on Make Certain from *Quality Is Free* because there is an interesting story about it and also because the program is very useful.

In the late 1970s service and administrative executives began to be interested in quality improvement. However, for some reason they were very sensitive to anything that was tainted by being associated with manufacturing. To them manufacturing was a dirty business full of smokestacks, bad labor relations, and cutthroat competition. I always asked these executives if they had ever been inside a plant, and I found that very few had. And none of them actually had worked in one.

The phrase that provided them with the most difficulty was "zero defects." They did not see their operations as being defective or nondefective. They felt also that ZD had been sullied by the Department of Defense's motivation program. They wanted something different that would provide a dignified way of asking people to do things right the first time.

While this process was emerging I had been searching for a way to bring more white-collar operations into the quality processes. In many companies they would remain aloof, not even participating in the search for problem identification. Yet these were the people who knew the operating systems better than anyone else. I kept trying to reach them in this regard. As part of this "reach" I began to talk with groups of people working in marketing, software development, insurance loss control, hotel management, catering, accounting, technical services, administration, and other service areas about their problems. They had a lot of complaints and felt to the most part that they were being poorly served by those who supplied them inside their own organization. What they received from other departments was usually incomplete and tainted. Oh, if only everyone were as good at their jobs as they were.

I didn't tell them that this echoed the manufacturing processes, since they seemed convinced that their position was unique. The key provision here was that they were interested in finding solutions to these problems as long as it didn't look like they were at fault. It occurred to me that we should deal with getting the processes to be correct rather than looking like we were working on the individuals. Out of this thought popped the phrase "make certain." I began to ask them what could be done to make certain that the processes and procedures they were using were error-free. This caused a positive response. Soon suggestions were coming in at a rapid rate from the operating people as well as their management. All of the suggestions were useful, some dramatically so. Ways of doing things better, worth large sums of money, began to emerge. I was asked to formalize Make Certain so it could be used all across the company and in supplier firms. The result of doing that is contained in this chapter.

The funny thing about Make Certain is that it absolutely worked in every case, but slowly faded from use after a few years. Zero Defects, which as we discussed is a performance standard, is still around. In fact it has become part of the language, I hear it all the time in the news and see it in articles about management.

I recommend using the Make Certain program as a suggestion vehicle in knowledge worker areas. They will like it and as a result will provide tons of improvements for very little expense. The following is the Make Certain chapter as it appeared in *Quality Is Free*.

INTRODUCTION

This chapter contains the Make Certain orientation concept and presentation on the following pages. Following that there is a complete step-by-step description of the events necessary in running a complete Make Certain program. It helps in the Quality Improvement Process.

Make Certain is a person-to-person, white-collar-oriented improvement program that gets everyone's attention immediately. You will receive prevention suggestions from over 90 percent of the people exposed to it. It is always difficult to get knowledgeable workers to identify problems and offer solutions. Make Certain has proven a valuable tool in these areas.

Use it with good luck. It really works.

INSTRUCTOR'S GUIDE FOR
THE *MAKE CERTAIN* ORIENTATION

TIME:

About one hour

EQUIPMENT REQUIRED:

Blackboard or other writing display material

AUDIENCE:

Fifteen to twenty-five personnel of white-collar or administrative functions. Preferably the attendees should represent many different departments or functions. However, the orientation can be given to people from one operation as long as the instructor is sensitive to special organizational or personality problems that might be involved within that function.

PURPOSE:

- To explain the concept of *Make Certain* in a way that will make the personnel involved want to participate in this program of defect prevention for administrative and functional activities

- To start an ongoing examination of procedures and methods by the personnel involved in order that they will contribute to defect prevention activities on a regular basis

SEQUENCE OF EVENTS:

1. Introduce the thought that many nonconformance problems are caused in the administrative, service, and similar activities of the company, and that they are long-range in effect.

2. Explain that *Make Certain* is a program to help identify those problems and eliminate them through soliciting ideas from the individuals involved in doing the actual work.

3. Ask each individual to state their personal "biggest problem" without detailed discussion. Write the problems on the blackboard.

4. After problems have been stated and written, comment that everyone has selected problems that are caused for them by others, that no one seems to have problems that they have caused for themselves. Note that this is a typical attitude among humans.

5. Ask participants how some of the problems listed on the blackboard might be prevented. Avoid embarrassing anyone or pinning someone down. Select one or two and give prevention ideas yourself.

6. Tell them how necessary it is for all of us to become "Certain Makers." Cite a few statistics showing the cost of error in "white-collar" areas.

7. Go around the room again and ask people to state their "biggest problem." This time the problems should be different.

8. Ask them to submit written ideas to the improvement team. Suggest that they might like to have their supervisor get together with them to set up defect prevention discussion teams in their operations.

9. Thank them for coming and dismiss the group.

MAKE CERTAIN MEETING

INSTRUCTOR SPEAKS:

Good morning. My name is _____. I am here to participate with you in a discussion of a new program called *Make Certain*. The purpose of this program is to help all of us who work with pencils, pens, computers, telephones, and other devices to learn more about our personal responsibility to quality.

As you know from your personal experiences many of the most frustrating and expensive problems we see today come from paperwork and similar communication devices. All of us have had problems with department store computers, catalogue companies, our internal departments, hotels, and other service functions that are supposed to make life easier for us.

Studies show that better than 25 percent of nonmanufacturing work is routinely done over before it is correct. Those are the jobs we are involved in every day.

The biggest single problem we face in doing our work is the communication that links our work together. Whatever your job is—management, computer programming, clerical, product line operations, sales, engineering, front office, accounting—any of these and all the others are bound together through a common need. This need is that we have to transmit our personal contribution to our jobs via pencils, pens, computer programs, conversation, or some specific determined method.

We receive data from someone. We decide something on the basis of that data, we transmit something along the line, and we add our two cents' worth to it. If we have not made certain about what we have done, then we can set the entire chain off in the wrong direction.

Business is a chain of paperwork and other communications that we control and utilize. The effectiveness of the business is determined by how well we do that data transmission.

Unfortunately it only takes one bad bit of data in the chain to disturb its effectiveness and accuracy. If we as individuals were electronic components, we would determine our communication reliability the way it is done with components. If you have 100 components in a circuit and each one is 99 percent perfect, the probability that the circuit will perform is only 35 percent. You have to multiply each individual reliability by the next and so on.

What we have to consider today is your individual reliability in this matter of making certain that we don't cause problems.

If it is possible to get every job done right the first time, then we will be able to reduce the amount of time we waste on rework, the number of customers we disappoint, and the amount of frustration we cause ourselves personally. We will be able to do more of the purposeful things we really like to do.

Instead of making up some typical problems as examples in order to make this point, I would like to ask each of you to state your biggest problem for us—the thing you regard as your biggest problem is getting your work done right the first time every time. I will go around the room and ask each of you to state that problem without discussion. I will write them on the blackboard, and in a few moments we will have a list of real-life items that we can discuss.

I think it is very important that we have a discussion based on something we would recognize as being pertinent to our situation rather than something that comes from outside our area.

NOTE: Point to each individual one at a time and ask them, "What is your biggest problem?" As they state it, make certain you understand it, and then write in on the board. Be very open and friendly. In this activity they have to understand that all of this is not going to be used against them.

Typically they will say things like:

- They don't send me accurate data.
- Management isn't clear about what they want.
- We can never find out when problems happen.
- They keep changing the standards.
- It is hard to get computer time when you need it.
- The salespeople wait until the last moment to send in the orders, then they want it immediately.
- The customers don't know what they want.
- There never is enough time.

Now we have all of these problems listed, and we can see that they have something in common. What they have in common is that none of them are problems we caused for ourselves. They are something that others are doing to us. This is the typical human reaction. It proves we are all normal, functioning human beings.

And of course it makes another point that we all have to recognize: Eliminating problems and improving personal reliability is not just a matter of concentrating and trying harder.

It is not just a matter of thinking how to be more careful.

That is like your new diet. It only works for a little while. Then you go back to your old ways. We all do that.

What we need is some systematic recognition of the basic problem. There are three specific recognitions involved:

- First, we have to recognize that the largest cause of defects and problems in any company is in the paperwork and other communication system areas. The factories have their own problems, but they are working with what we give them.

- Second, we have to recognize that every problem is preventable and that the person who can best contribute the idea to prevent it is the one who has at some time caused it or a similar problem.

- Third, we have to recognize that although we are hearing these words and agreeing with them, we as individuals do not really believe that they apply to us personally. That is only human.

The way to get started on making certain is to recognize that we cause problems for ourselves, and we must find ways to prevent them.

Closing Sequence:

1. Review the "biggest problems" for possible solutions.

Let's look again at these problems we put on the blackboard. I don't remember which ones belonged to whom but let's take a couple of them and see how they might be prevented if we can get the right people interested. For example:

- If management is not making the instructions clear, write a procedure and get management to agree with it or change it to what they want. Then it will be clear.
- If a group fouls up what you send them, perhaps you need to think out how to explain it to them in a different way.

These kinds of ideas are well within your experience.

NOTE: After you have given a few, encourage the audience to contribute some.

2. Give them an idea submission format so they can go back and come up with some ideas to make their life easier.

3. Ask again, without writing on the board: "What's your biggest problem?" This time you should get many people saying that they have to get in there and make a contribution to getting things done right the first time. They are going to recognize that they are part of the problem. Suggest that they might like to have regular meetings in their groups in order to discuss defect prevention on a regular basis, like corrective action groups in manufacturing.

4. Thank the group and dismiss them. If anyone wants a fuller discussion or wants specific directions, ask them to stay after the meeting in order that the group might move along.

EVENTS IN CONDUCTING THE *MAKE CERTAIN* PROGRAM

1. Brief management staff on the program concept and intent. Agree to appoint coordinators for each department. Remember, the emphasis is on paperwork and service operations.

2. Meet with coordinators to explain the program. Ask them to plan a meeting of all the supervisors in their department in order to orient them to the program. Solicit examples of cases where defect prevention or more attention to detail could have saved problems and money. Tell them they will need at least three of these examples. They don't need to be big things. For instance: billing errors where the customer returned the bill because something was wrong with it, thus we didn't get paid on time; purchase order errors where the wrong information was given, causing the wrong product to be delivered; work instruction errors; and the many other things that supervisors complain about when they discuss their work. With a little encouragement, and the assurance that it will not be used to embarrass them, the examples will flow.

3. Depending on the size of the company, the supervisors will be taught how to use the program by the coordinator, if there are not too many. Otherwise the coordinator must teach the department representative who will then instruct the supervision, including the department head.

4. Supervisors meet with their people. They explain the logic and concept of the program and tell why the company needs their help, citing some examples. Then they have this discussion.

   - Who is our customer inside the company? (It could be another department, the president, or whoever receives the result of our work.)
   - What specifically does that customer want from us?
   - What could we do, specifically, to make sure the customer does *not* get it?
   - Who is the ultimate customer, the one who uses the product or service or the company?
   - What specifically does that customer want from us?
   - How can we make certain that the customer does *not* get it? (Keep all this light, but insist that the question be thought out.)
   - Select the best actions and discuss them. Then ask everyone to suggest how we could measure these items. For instance: if we were to always send the bills to the wrong address, we would know we had failed to do wrong if one of the bills didn't come back. That would mean we had made a mistake and sent it to the right address.
   - Write down the suggested measurements because they will form the basis for positive action measurement later.

5. Ask the individuals to turn these thoughts around and make suggestions of ways we could make certain these failures *don't* happen. The form says: "How can we make certain that our customer is receiving what we have said we would provide?"

6. Receive the suggestions and give a "Certain Maker" badge for each one submitted. The supervisor makes his or her comments on the suggestion card and sends it to the coordinator.

7. Make a lot of fuss about suggestions that are implemented right away. Post ideas and photos on bulletin boards.

8. Measurement of the suggestions should be based on how many hours of work do not have to be spent as a result of something being done right. Sometimes material will be involved also. Multiply the hours by the loaded average wage to get a dollar figure. But remember that the idea of the program is to instill the attitude of defect prevention.

9. During the last week, have a countdown: four days left to Make Certain, three days left....

10. At the end of the program, write a letter to all employees, thanking them for their participation and explaining that no suggestions will be left unscrutinized.

11. Implement as many suggestions as you can, and keep reporting on that for the next several months.

12. Study and learn the orientation program. For best results, do it exactly the way it is written.

EPILOGUE

The chapters that precede this Epilogue cover over 40 years of an experience that could not have been projected by looking into my crib. I feel very fortunate to have lived it. But what about the future? No one ever knows for certain, of course, but certain patterns do expose themselves. It does all sort of come around again.

In quality management, during this experience time, we have gone from "you make it; I check it; she fixes it; and keep your trap shut" to a fixation on getting things right the first time. Management at all levels has realized that giving the customer what they ordered is no longer an option, or a big deal. If you do that you have no advantage, others are doing the same. If you don't do it then you cannot participate in the game. It is still hard work, causing quality to be routine, but most folks are working on that objective. What they are doing is not always productive because it is often based on procedures rather than comprehension. If you wish to see how this works watch the early morning infomercials promoting golf clubs that will make your game successful. But these things take care of themselves, what matters is that they are interested in improvement.

During the past three years I have received and accepted invitations to speak all over the world: Brazil, Mexico, Chile, Venezuela, Jamaica, Australia, Singapore, Malaysia, Japan, China, India, Greece, United Arab Emirates, Saudi Arabia, Spain, United Kingdom, Canada, and a few other places as well. I have turned down a bunch due to lack of time or resources.

The thing all these sessions had in common was that the groups wanted to talk about being competitive. This boils down to "How do we sell our services and products all over the world, and build an economy on that?" They had accepted my Absolutes of Quality Management as being a given. They did not want to argue about the wisdom of trying to get zero defects, they knew that was what had to happen. What they wanted was to know how to get all this done. I told them that it lay entirely in the hands of management and used the PCA case as an example. It was during these sessions that I learned to describe the target as "complete transactions and successful relationships." That they understood, which is why I made it the subtitle of this book.

The hand-to-hand combat of the past 40 years about what quality is and how to get it is past. Everyone agrees that quality is a vital part of running companies that produce products and services. They often differ on exactly how to go about it, but that is good for our learning ability.

The future is rearing its head now in the area of information. Here we are back to the caveperson days of quality management philosophy again. No respectable software developer expects to put together a Zero Defects product. They have the same patronizing smiles for me that automobile, bank, hotel, health care, and other people used to provide. They will get away with this for a while. But this will change as people begin to really understand what the information age is all about. It is not just playing games on computers, it will involve every tiny part of our lives. If it is not done properly then we will suffer. Right now people can deal with a little lack of integrity in their information. Backup systems exist to provide a checking capability. But as it all moves quickly forward we will be dealing in personal matters worldwide without even knowing it: bills will be paid automatically as directed in our programs; pocket cash will disappear to be replaced by a card or token; each of us will have a worldwide telephone number; medical diagnoses will be conducted by wireless transmission; automobiles will contain hands-free driving capability; maps will commonly be on display screens; electric power will be delivered by satellite. In short it will be a world where the accuracy of data will determine the majority of our safety, happiness, and activities.

Systems integrity will be the concern rather than just quality, as we see it today. If everything is not accurate and reliable, then there will be no integrity. If there is no integrity then the information age cannot evolve. If it does not evolve then the world economy will not come close to reaching its potential. Life on our planet will not improve, and probably will not survive as we know it.

The next great crusade in quality is to convince those exceedingly bright folks who run the companies and organizations dealing with information that error is not inevitable, that Zero Defects is a desirable and attainable management standard, that nonintegrity costs them an incredible amount of money, and that the way to deal with problems is to prevent them.

Sounds familiar doesn't it?

Go get 'em!

GUIDELINES FOR BROWSERS

People must learn to react to what is going on; they must have a base from which to function. 3

They do not have time to remember and apply the systems. 3

Quality does not come about as the result of some particular way of dancing. 3

People have to be helped to understand that they reside in a culture where the correct action and result are desired. 3

The way to prevent fire fighting is to not have fires. 3

The idea of causing quality to become a normal part of an organization's operating arsenal did not catch on automatically. 3

Going through the Baldrige criteria, the ISO series, or some similar list of activities results only in books of procedures and wasted efforts in urging people to comply with them. 3

All the automobile drivers in the world are "Certified," which has nothing to do with their capability of driving or their system for doing it. 4

Rather than coming from systems, quality actually emerges from the way management presents policy, education, and their personal example. 4

Mil-Q-9858, birthed by the United States Department of Defense (DOD), has been around for 40 years and there is not a single case I know of where it caused a company to produce conforming material and services routinely. 4

Quality had always been conducted as though it were a difficult task in organizations, whether they considered their business to be service or manufacturing. 4

Prevention, as a work ethic and practice, must be deliberately inserted into the operating culture of an organization if it was to have an effect. 5

This divergence of concern between the conventional wisdom of quality and myself began when I proposed the performance standard of Zero Defects back in 1961. 5

Waste was routine. 5

Automobile companies were spending at least $2000 a car just on repair during assembly, plus warranty and recall expenses after the sale. 5

Simply, we just had to learn how to get the right things done right, the first time. 5

To me it was only common sense; others did not see it that way. 6

Today most company management are still struggling with quality, expending a lot of effort on systems and activities while achieving small results. 6

The organizations with established quality-control departments were working on getting better, but the responsibility was left entirely to that function. 37

I had been complaining about the American attitude toward quality for 20 years and receiving little positive response. 38

American managers and their quality professionals had been hung up on the idea that it cost more to make things right. 38

Many people confuse activity with results. 38

There is nothing complicated about the concepts of quality management. Causing them to become part of a company's culture is not difficult or expensive. 38

Uncertainty has always seemed to me to be like something freshmen undergo when planning for college. 39

This ["let's do it too"] is an idea-of-the-month mentality where executives look at the surface idea and launch an effort to show that they are up-to-date. 39

"The Japanese quality model," for instance, exists mainly in the imagination of the media. 39

During this stage, the common thought is that doing things right means doing them slowly, involves a lot of checking, and requires a reduction in innovation, productivity, and creativity. 39

The CEO picks up the criteria for the Baldrige or NASA or some similar award, hands it to someone, and instructs that person to "go do this." 40

If no lawsuits or horse whippings are forthcoming, they assume that everything is in good shape. 40

If asked to state their performance standard, the employees gaze blankly. 41

The reason things go on like this is that management is not interested in numbers that do not have dollar signs in front of them. 41

After the first half-hearted swipe at quality improvement goes nowhere, management stirs itself to look for some external happenings that will make a difference. 41

The Wizard of Oz solved the Scarecrow's problem of having no brain by giving him a diploma. 41

It is like finding a doctor who fixes the X rays rather than doing surgery on you. 41

Imagine yourself standing on a box in front of 500 employees and instructing them to go out and "delight the customer." 42

Setting the employees in a conference room and showing them a few hours of tapes changes nothing. 42

We run our lives on concepts, no computer program can handle a marriage, nor can a set of processes deal with our budget. 73

In quality management these are the Absolutes: Quality means conformance to requirements; Quality is obtained through prevention; Quality has a performance standard of zero defects; Quality is measured by the price of nonconformance. 74

We have to understand the difference between guesswork, false dogma, and reality. 75

When I tested the conventional wisdom of quality that was preached to me, I found that it was not in touch with reality. 75

Apparently being found competent in one thing gives you some credibility for being competent in other areas. 76

It was becoming apparent to me that the way things were done was not very efficient. 76

Life is a lot easier if you deal in mythology. 77

I had learned to keep my radical ideas to myself while plotting to overthrow the government of quality. 78

My main work was to look at the nonconformances that the inspectors and testers had identified and then classify them as to seriousness, cause, and responsibility. 78

They had a system for sending things back to suppliers, but to his knowledge nothing had been sent back in months. 79

The supplier never knew anything was wrong and kept doing it that way, and purchasing never heard about it. 79

This was when I began to realize that we had to establish exactly what the word "quality" meant if we were to get things done right. 79

We had to have a concept of "conformance to requirements" if we were going to learn how to have a smooth, efficient work flow. 80

I developed the concept of Zero Defects (ZD) because we could just not learn how to find everything that could be wrong in a weapons system. We had to prevent, not sort. 80

The most important thing I learned at Martin had to do with relationships. 81

It only takes 15 minutes to explain that happy and successful work comes from learning how to do the job right the first time. 81

The bigger their offices, and the broader their privileges, the less they knew about reality. 82

For some reason when people reached that position of authority they became antagonistic to the levels they had previously occupied. 82

The concept of treating everyone the same is valuable. It is the best defense against becoming arrogant, which is the fatal disease of executives. 83

What I had to do was create an organizational culture that understood and practiced the concepts of quality management as I had conceived them over the years.

87

First we had to create the playing field by establishing clear policies and practices; second, the quality professionals, the corporate executives, and the operating management needed to be educated.

88

Most of the people I met were interested in having me get the other people on the right track.

88

The quality policy of the ITT Corporation is that we will deliver products and services to our customers and co-workers that meet the agreed requirements, or we will have the requirements officially changed to what we and the customer really need.

90

I discovered a little booklet written by the CEO. It laid out a way of writing a memo and was titled "Unshakable Facts."

91

In ITT the staffs had the reputation of being very tough on the unit people. The old story, which I stole somewhere and have used for years, is that they are called seagulls: "they fly in, squawk a lot, eat your food, crap all over you, and then fly away."

92

No one paid much attention to the quality manager in the "commercial" world unless they had just the right kind of personality, and none of them did.

94

The quality function people would become very important and useful if they cooperated with me and learned how to get the right things done. Lack of cooperation could produce the opposite result.

95

I did promise that every place I went I would solve at least one current problem and save them enough money to pay for my trip.

95

Personnel was a thorn in my side for the next 14 years. It wasn't that they were picking on me; they just had their own agenda.

96

They did not relate to the organization or its culture; they lived on their own planet.

96

Asking for permission to do something is not the way to get things done in a large corporation. No one with authority will ever want to do anything different.

97

However, the things you are doing have to be presented one day and that has to be done in a manner that does not appear to threaten anyone or any existing programs.

97

It takes a great deal of selling to convince people that your intention is to just make life easier and more productive for them, without adding any control over them. They are not used to someone wanting to be useful.

97

There was little contact between those with suits and those wearing work outfits.

97

They were glad to see me but they would have welcomed Hitler back if they thought it would get Brussels off their backs.   98

The Europeans like specifications and requirements; they never seem to get enough of them. If none exist, then they will create a bunch. Once created, they never stop existing.   99

If you can keep the subject of quality aimed at paper instead of products or service, it lowers the level of trouble.   100

The quality systems inside these companies were patterned after the quality-control methods developed in Western Electric. These were aimed at detection and correction.   100

I had learned that the worst way to initiate change was to do it by fiat from headquarters.   100

I learned to bring the Europeans over to the Western Hemisphere and let them lecture the natives. That had a positive effect.   101

I wrote a booklet called "Quality Improvement through Defect Prevention." It covered the concepts of quality management, plus the 14 steps of an improvement process.   101

We had to lead people away from the irrelevance of Quality Control and Statistical Process Control which were being taught in the profession.   101

I finally learned to just go ahead and get what I needed. No one seemed to care—when we are dealing with billions, a few thousand is not much to worry about.   102

Companies spend a lot of time and money controlling small amounts and terrorizing their people in the process.   102

Geneen liked to hire bright people with senior executive experience and then just sort of toss them into the pits to see what they found to do.   102

My sources, through the quality council and my personal relationship with the quality managers, let me know what was really going on in the unit, good and bad.   103

All of this effort led people to think that they had better things to do than make trouble for the quality function.   103

It was all right to have problems; it was not all right to ignore them or not ask for help.   104

When executives talk about quality problems they mean that something didn't perform correctly. They do not mean that the Quality Department itself was the culprit, but other people think that is what they mean.   105

In the process, it is necessary to make certain that individuals are not targeted. "Fight the problem, not the people," said Hal Geneen, and he was correct.   105

Over the years I have developed the theory that there are actually only a dozen fruitcakes in existence, they just keep being sent around. We used ours as a doorstop.

We were the smallest staff in headquarters. But we became the best known.

It is a cliché to say that business executives are usually not good speakers. It is also true.

They become adequate and avoid embarrassment, but success eludes them because they never really thought they had a problem in the first place.

Instead of arguing about quality, or blaming the quality function for problems, the executives understood their personal role and insisted on a standard of Zero Defects.

The Ring of Quality program was working well worldwide. People were asked to nominate someone, other than their boss, as the person they recognized as their personal example of quality performance.

Systemwide, things were much better in the matter of quality now; everyone was aware of it and took their personal responsibility for action.

The role model of the resourceful manager who could deliver regardless of the situation began to vanish. Senior management started to appreciate those who did the job properly without a lot of fuss and bother.

They did not want to be snatching victory from the jaws of defeat all the time; they wanted a smooth journey. It was less painful and made a lot more money. It was also easier to manage.

These nonmanufacturing companies did not immediately recognize that they spent about half of their effort doing things over.

What the hotels looked at as gracious attention to the guest was actually rework.

Insurance companies rarely seemed to get the policy written right the first time.

The education of the executives who run those kinds of companies has some blank spots in it when it comes to understanding what is involved in actual work.

I had learned to teach them to do "elimination," not "reduction."

If we could encourage people to look at their job and take a dollar a day out of it, then we could have a large cost elimination with little effort.

I had a heart attack early in 1972 and learned that I was not immortal after all. This shocked everyone, including me, but it did convince me to stop smoking and to begin to take wellness seriously.

GUIDELINES FOR BROWSERS

I could tell I was beginning to get bored with the challenge that remained. Still a lot of interesting things were happening. 117

It is very difficult for a young tiger to emerge in a structured organization. What I was doing was successful only because it did not threaten anyone or step on any ground with footprints already planted. 118

I worked hard also, but kept regular hours, stating firmly that my family came before the company. 118

You can get to thinking that you are doing a lot if you put in long hours. However, much of the time executives in all companies concentrate on things that make little difference. Exhaustion is not always an indication of results. 119

They all asked if this was my first visit to Japan. Since my previous one had been as part of the occupation forces, I usually lied about it and said yes. 119

The main impression I received was that the general managers explained their quality process and plans to me rather than delegating that to the quality manager as in the United States and Europe. 119

They thought that my ideas on quality management were very advanced, and knew that I was not very appreciated by my colleagues in the United States. 119

The market was all the Asian countries, but the Aussies did not like Asians much and so did not build up relationships. 120

I realized from this two weeks of exposure that I was going to have to include these areas in my thought process if I ever started my own education firm. 120

It began to look more like my business would be education rather than standing around offering advice. Everyone needed to learn. 120

I did not want to become executive vice president or even president of the corporation. I thought that these jobs were mostly reacting to the problems other people caused and had little to do with real leadership. 122

I began to feel unappreciated again, which is the proper process if one is to determine to make a change. 122

I had no desire to go building to building asking companies to be my clients. The problem was to figure a way to have them come to me. 122

The answer had to be a book that executives would read. 122

In the years between writing my quality management books, I had been an actual executive, living with others, and answering for any crimes that were committed. 123

Those who wrote and taught about quality management, quality control, quality assurance, reliability, and all the other titled functions usually had no high level operating experience. 123

American executives were coming under the impression that the problem of quality in the country was the American worker. In reality it was the management who had become separated from both worker and customer through years of success. 123

Textile manufacturers, faced with defect-free carpet from overseas, still insisted that 15 defects per hundred yards was the proper standard. They pushed for government regulation in order to stop others from being better at quality. 123

The quality spokespeople were saying that the need was to work harder at quality control. But that was not getting anyone anywhere; it just made the products more expensive. 123

So I wrote *Quality Is Free: The Art of Making Quality Certain.* McGraw-Hill hated the title, and looked fruitlessly all through the material for some charts or quality-control stuff. 124

I found no interest among the majority of board members in getting the country straight on quality. 124

Most executives are so self-centered that they do little to grow their people. 126

I decided that it would be run in a manner that would make it a wonderful place to work, and at the same time it would be more profitable than any other organization in the field. 126

We will take the creation and accomplishment of requirements seriously. 126

All employees will be called associates, will be treated with respect, will be selected carefully, and will be given every opportunity for professional growth. 126

We will teach those who instruct in the Quality College, or consult with our clients, to understand the Absolutes of Quality Management. We will hire none with quality-control experience. 126

This will be a management education company, not a consulting organization. 126

There will be no Personnel Department. 127

We will have a pension program, thrift plan, and a fair hospitalization supplier. 127

Management will be required to keep in touch with the reality of daily life. 127

We will have a monthly family council meeting where everything will be open for discussion. 127

The executives for whom I wrote the book actually related to it and were reading it. 129

Quality-control professionals, particularly those who were prominent consultants and teachers, dismissed it quickly. One said "The clown has written a book." 129

Some dismissed my work as being "worker motivation" or "exhorting the workers." I never understood this attitude because there is nothing about that subject in this book, or any of my other ones. 129

There was a general understanding, deeply embedded, that quality was varying degrees of goodness and that the more good you got the more it cost. 130

They were fixed into strategy planning, financial management, return on investment, and the other tools of conventional industrial management. 131

They often said that if they could run their factories with robots, that would be their preference. 131

This attitude, which turned out to be very destructive, is what I call "mainframe thinking." 131

It became possible for me to know early if a specific company could change its thinking and its culture in relationship to quality. 131

I often feel that something happens to people's hearing equipment when people hit the top levels of an organization. 131

There were two types of companies in those days: the arrogant and the searching. 132

However, those who had a big training department or quality function usually decided that they could do this themselves. It all sounded so easy. 132

As a result they wasted many years trying to change the people rather than managing differently. 132

They worked on the wrong stuff for several years because they were too proud to let someone show them. 132

My experience in business had been so broad that there was very little that I had not been through before. 133

One of the most amazing people I ever met called one day, when the College was about a year and a half old, to say he had read *Quality Is Free* and wanted to start his company down that road. This was Roger Milliken, the CEO and owner of Milliken Company, the textile firm. 133

It was a joy to see a group that didn't want to argue about it; they just wanted quality to be free. 133

Concerning clients, I have always been very proud of the fact that we did no sales work, never made a call—they all came on their own. 134

Six thousand executives and managers a year were attending the Quality College before long and many thousands more people were utilizing the tapes and other internal material we developed and taught. 134

If this was to be a real company one day, then it should be designed so that people would be proud to work there and clients would receive the most help they could stand.   135

During this strategy phase I ran across a comment by Warren Buffett describing a "wonderful business."   135

Taking all this with what I had learned over the past few years, I wrote out a strategy at the end of 1979.   136

The management had to understand that they were the cause of the problem, and the employees of the company all had to understand quality the same way.   136

What was important were the concepts, not the techniques.   136

Statistical process control, Quality Circles, and other popular programs had nothing to do with the cause and effect of quality. They were just tools, and properly applied could be useful.   136

I built the Quality College around the Four Absolutes of Quality Management as they had evolved for me over the years.   136

Quality means conformance to requirements, not goodness.   136

Quality comes from prevention, not detection.   136

Quality performance standard is Zero Defects, not Acceptable Quality Levels.   136

Quality is measured by the Price of Nonconformance, not by indexes. 136

We held our first management college class there in October of 1979, and then did one in November, and December.   137

The Lord always provided just the right person for PCA at all levels of the company.   138

One of the advantages of our location was that we could walk our students to lunch on Park Avenue. This let them get exercise and drink in a little local character.   138

We were able to produce Zero Defect lunches on schedule.   138

The students (executives and managers mostly) came to the Quality College with the idea that if we were going to teach them that an operation could be run properly, then we should be a living example of just that.   138

Each associate was selected for employment based on their desire to get everything done right the first time.   139

My first priority for the company was that each associate would be carefully selected as a result of his or her personal commitment to really contribute and that we would all treat one another, as well as the clients, like ladies and gentlemen.   139

Senior people in business just weren't very nice to the lower-level employees, at least that had been my experience.   139

It is a very rare management that will voluntarily take the time and effort to conduct such communications. 139

I selected people to be instructors based on their experience in real-life management and excluded anyone with a quality-control background. 140

In September of 1980 we went back to Greenbrier for our physicals and Dr. Morehouse reported that there had been a significant change in my cardiogram during the stress test. 144

The upshot was that I went to the hospital and had a heart catheterization examination, followed by an operation in which five bypasses were made of the coronary arteries. 144

It is a mark of honor for many professional people to have papers piled up on their credenzas and desks, but it looked bad to the clients. 145

We were getting a steady stream of companies wanting to come and talk about their strategy. 145

Some of the more arrogant companies looked down their noses at actually being taught communications. 147

I established the "Beacon of Quality" award for all hands, and the instructor certification plaque for that group. 147

Around June the recession began to hit corporations and, true to form, they immediately began to reduce what they were spending on education and training. 149

My suggestion was that instead of laying off 20 percent of the people, we would all take a 20 percent pay cut. 149

One was about the Flypaper Company who spend most of their money going around the world looking for flies who were weaker than the glue they were making. 151

People have their own agenda and most of those schedules are not too productive. 151

We needed a company, not a gathering. 152

Since QES represented a new concept in quality education, the clients did not absorb the idea immediately. 153

They could do the "teaching" and the material would assure that everyone learned the same thing. 153

Professionals become self-centered in every business. 153

We found that some of the training people in client companies would duplicate the notebooks or cover the pages with plastic and use them over again. 153

We were still struggling with the Barnett Bank people who were very worried about their money, although we had not missed an interest payment. 154

GM offered to purchase an equity position in our company in order to make things more permanent. 155

It is harder to manage success than tribulation; people don't listen well when it comes to dividing the spoils; they do when we are launching lifeboats. 155

The meetings opened with a prayer and everyone was nice to each other. We fought the problems, not the people. 156

People do have their own agenda. Usually they do not see the future past their own commitments. 156

No one looks at the whole organization except the CEO. 156

My experience as a consultant and teacher for the past four years had shown that management really didn't like their employees very much. 157

They honestly didn't seem to understand that the people were their primary asset. 157

The main emphasis of QWT was to spell out the concepts in clearer detail, the "Four Absolutes" as I called them. If one could understand these, then the whole business of quality would be revealed. 158

Management is so technique-oriented that it is hard to get ideas through to them. 158

We had thought out the procedures and policies and incorporated them into a strategy that everyone could understand. There were no sudden changes; everything was planned. So the panic calls and noncompleted assignments usual in most companies did not exist in PCA. 158

I established a "Thanksgiving Week" that would occur in April each year. The theme was to thank all of those who had made it possible for all of us to be living so well. 159

It is very difficult to teach people about things they have not experienced. 160

It also made me wonder about all those who were conducting seminars on quality management who had never actually managed any quality. 160

We started annual Alumni Conferences in order to help clients bring each other up-to-date. All the speakers were client people talking about their experiences, and the results were wonderful. 161

What we taught was a culture change which required management commitment, education, and action. What most executives wanted was a series of actions that could be delegated to a functional department like Quality or Human Resources. 161

One company program that was being well managed was contributions to charities. 162

We said that we wanted to give money to "real people helping real people." It is not as easy as it sounds. 162

Now and then someone would indicate that we should share this with the shareholders rather than giving it away. My response was that God had made us what we were for this purpose and that we needed to honor our commitment to Him. If they wanted to do it differently, I suggested that they go start their own company.

163

I thought we should consider going public. No consulting firm had done that before.

163

The College classes were conducted by people who had been taught how to do it, not by professors or individuals doing their own thing.

163

"Due diligence" means that everything one says about an organization must be verified so that the person considering a purchase of shares will be told the complete truth.

164

My reasons for wanting to go public were really three: first, I thought the company needed that sound base of being able to reach out for funds and never suffer the indignities that were placed on us by our Bank; second, the associates and family members who owned stock would now be on their own, and I would not feel responsible for them having stock that was not marketable; and third, I wanted our family to have money so they could be independent.

164

By the end of 1985 PCA had sent almost 12,000 executives and managers through the Quality College.

165

Dealing with people in the most agreeable of environments with the most enlightened of managements is still the most challenging of tasks.

167

It is hard to get people to treat each other like ladies and gentlemen. If people did, Shakespeare would not have had anything to write about.

168

Writing a book is a fine involvement for me. Everything comes out of my head, so I don't have to plow through other books to lift and footnote quotes from other writers.

168

PCA was a happy company, the people were proud to work there; it was profitable; the customers liked it because it was so professional. All of this came about because of actions I had taken, or programs I had installed *on purpose.*

168

We had a quality improvement team, for instance, and appointed a new one each year. They did a lot of good work, but their efforts would have been for nothing if the culture was not one that wanted to improve.

168

Every person who came to the company on a visit remarked within their first half hour that this place was different.

169

The executives and managers who were our students did not recognize at first where all this came from.

169

I realized that the company was going to change after I left and also that I was getting ready to be gone.

169

My mentality had always been to create something, develop it, show others how to do it, and then move on to something else. 169

I can't stand conflict, the kind where people are unkind to each other, or yell, and argue. 170

For that reason I went out of my way to make certain that the environment of the company was peaceful and purposeful. 170

Keeping quality installed in a company is a full-time job for the senior executive. 170

Client companies wanted to know about case histories, what had happened to other companies who had implemented the quality concepts and processes we taught them. 171

Actually no one ever does anything much differently after being exposed to a case; they immediately say that it does not apply to their specific situation. 171

Being human, they were not necessarily interested in improving their company, the real concern might have lain more on the way their own efforts would be viewed. 172

When each new associate was being oriented, and we included those client people who came for extended instruction also, they had lunch with me for the ADEPT presentation. 173

A is for accurate—we do what we said we were going to do and when we give information we know it is right. 173

D is for discreet—we do not gossip, and we treat everyone like ladies and gentlemen. 173

E is for enthusiastic—we try to hire enthusiastic people and then not turn them off. 173

P is for productive—the more of us there are, the less there is to share. 173

T is for thrifty—we do everything first class but there is no virtue in throwing money away. 173

People asked me how I could stand to do all these talks on essentially the same subject, and I noted that the audience was different each time. 174

The key to successful speech making lies only partially with the speaker. 175

My experience has been that left to themselves the arranging committee will put the speaker behind a lectern, hemmed in by others sitting on the podium. 175

In the same manner visual aids are not friendly to the speaker. No matter how well thought out or attractive the viewgraphs are, they detract from the speaker's message. 175

Sending people off in groups to do some workshop or other is not useful as far as acquiring knowledge is concerned. However it does use up time and lets the people get some exercise while the speaker takes a rest.

When one is the chairman, founder, and chief guru of a company it is not easy to acquire constructive criticism.

People have to die one day, but if they take care of themselves they can keep from rushing into it.

Corporations do not have to die, but they do, committing suicide, as a matter of fact, in most cases.

I often receive criticism from reviewers and those who write about quality in particular, stating that my ideas and processes are too simple, that nothing really works out that way.

They are not looking for a philosophy based on ideas; they are looking for a regime based on techniques and procedures.

The ones who respond to me are executives who have had the problem of trying to lead some effort where there were no paths.

But it is up to the person to determine how healthy she or he is going to be. Corporations have this same problem.

The good news part of being an author is that people seek you out wherever you are in order to share their feelings about your products.

For PCA to grow worldwide, it seemed that we were going to need a partner.

We daydreamed about finding a partner who was already in places we were not. They could introduce us around and let us use their phone.

I was becoming more interested in writing and speaking than in running the company.

At this time I was writing *Leading: The Art of Becoming an Executive*. Still trying to reach the unreachable, I put this in the form of a novel with a leading character and a recognizable story.

The book also had a new and practical philosophy about the focus of a leader: finance, quality, and relationships.

Quality was portrayed as the structure, the body of the organization; Finance is the blood supply; and Relationships the soul.

We received many speaking invitations, and although they were interested in quality they seemed to just want to know what I was thinking about.

That was an interesting change from the past; they actually were becoming interested in me personally, not just in the nuts and bolts of quality.

I found that many senior management teams were concerned that they were not getting anywhere with the quality programs they were conducting. 183

In these sessions we look at the company as if it had no people or buildings, just a green field. Then we put on the field what we really need to run the company. 183

The evolvement of the Quality Management concepts in my mind during my career is something that did not happen to other people. 184

The content of my books is built around stories that came from the real-life involvement I experienced. 184

People who write college books about quality always solemnly put the list (14 steps) in a box, doing the same with other people's lists, as if that were the entire content of what we are thinking about. 186

Many folks looked at the list the way they do a scavenger hunt. 186

Very little thought was expended as to what the list was about in the first place or the effect it was designed to obtain. 186

When people really want something personally they figure a way of making it happen. 186

The same is true in political history where the idea is to gain and keep power. 186

What happens with quality improvement as a process is that the overall reason for doing it often gets lost somewhere early in the trip. 186

Instead of searching out a future that is bright, prosperous, and free of pain, quality management people get all involved with examining, documenting, and carefully placing the stones that cross the stream. 187

The more detailed information you give people, the less they accomplish. 187

The idea of having a formal process, in anything, is to take advantage of the thinking that has gone on before. 187

Quality is something that always starts out to be easy. 187

Often the one who is the biggest problem appears to be the biggest helper. 188

Not everyone really wants to get everything done right the first time. 188

If it makes so much sense to get the right thing done right, then how come the normal way of operating almost everything is to do the wrong thing first and the right thing second? 188

Just having a team doesn't mean they can play whatever game comes up. 189

It was never my intention that the team should run an improvement program; I was just talking about communication. 189

This objective is reached by having "quality" placed as the first item
on the monthly management meeting.
190

A permanent measurement system comes from setting up a Complete
Transaction Rating.
190

When we call it the Price of Nonconformance we have now defused
it. Every transaction in the company is accomplished either in a conforming or nonconforming manner.
191

The purpose of quality awareness is to let everyone feel that they, as
individuals, belong to this new company attitude.
191

There is no need to attempt to convince through awareness. That
comes from education and experience.
192

The idea of a formal corrective action system is to get everyone in the
habit of solving and preventing rather than learning to live with
problems.
192

There is nothing motivational about Zero Defects; it is just plain old
specific communications.
193

This committee needs to arrange a communication between those
who establish the performance standard and those who perform it.
193

The real purpose of ZD planning is to place management in a position
they cannot wiggle out of. If they are permitted to change in and out
according to the stresses of the day, ZD will be a myth.
193

Supervision, at all levels, from the chairperson down, seems to be
brain dead when it comes to passing along performance standards or
job understanding.
193

Most of the quality education classes are about things that make management think something good is happening. People need serious information that relates to the way they make their living.
194

Education and training must be accomplished in a planned fashion
and be conducted relentlessly.
194

The idea of having a celebration day was born to implant the message
that things were different around here.
194

There are just enough interesting things to show that this is a different
day and help people feel happy about their work. It is also a good time
to present awards, peer nominated of course.
194

Some people make fun of having a celebration day because they think
that it is the whole ball of wax. Those same folks go happily to weddings, christenings, recognition dinners, inaugurations, and such.
195

People often consider the practice of goal setting to be some sort of
ceremony, but it is a normal part of everyday life and conversation.
195

The best way to work on goals is to incorporate them into personnel
reviews.
195

Evaluation should be based on the accomplishment of agreed, and clearly measurable, goals.

The idea of error cause removal (ECR) is to help individual employees cause corrective action that affects their work.

It is essential to remember that the ECR system is for communication and that the individual must be included in the cycle.

Awards should be meaningful. They shouldn't be some form of money; they don't have to be valuable.

Military medals are trinkets but well respected because of the way they are presented and treated. People do not nominate themselves for such things.

This is my primary problem with government quality awards such as the Baldrige. If you have to nominate yourself, and even pay for an evaluation, what worth can it be?

I learned, once again, that people work enthusiastically when they feel they are a vital part of the operation, and they are slothful when work is imposed on them.

The wise leader helps arrange their agenda in a quiet fashion.

Selecting a lifestyle that produces the desired results requires eternal vigilance and action.

Driving to Grandma's house requires making a certain amount of travel and turning. If one movement is omitted it means that a great deal of effort and gasoline will be expended for inadequate results.

Companies, like people, are continually changing feet of clay.

My analysis is that there is no "wonderful" company. There are many that have been successful for a while but few who have maintained that status.

We should not wrap ourselves up in things that were well known, or that had survived the "test of time."

Hanging around doesn't necessarily mean something is correct or useful. People have often had the wrong beliefs for thousands of years, and have defended them with their swords.

We have to look at everything in today's terms and decide how realistic and useful it is right now.

If we are going to exercise by walking or running we have to have a philosophy of doing it that brings us back to our starting point when the trip is over.

People get heart disease for a very short list of reasons. The leading cause of lung cancer is smoking, just one simple item. Overweight and stress bring high blood pressure about. If you spend more than you make, you will have financial problems. The reasons are simple but overcoming them takes policy, education, and hard work. 206

The portrait of a tough company:

- Finance is focused on results.
- Management knows the customer.
- Employees are veterans with a high CTR.
- Distractions are avoided.
- Management understands a common agenda. 207

Those who build comfortable offices and building will find that it is difficult to get people to leave them in order to go where the action is taking place. 208

Companies that calculate and announce PONC set in motion a great deal of preventative action. This is not just platitude; it really has an effect. 208

Focus on the result. 208

Those companies that have lost their way usually do it by becoming separated from their customer. 209

They tend to think of the customer as being one large homogeneous group, all acting and dressing alike. 209

Managers who want to keep their jobs and their track record have to literally get out and rub elbows or tummies or something with those who actually purchase their product or service. 209

In order to acquire veterans who will have high Complete Transaction Ratings, it is necessary to take certain steps. These are selection, training, encouragement, and positive relationships. 210

The problem with selection is that most management does not recognize talent when they see it and most Human Resources organizations would not be able to pick Einstein out of a lineup. 210

Training is not a one-shot item. The content of work changes continually; office technology is always on the rise. 211

Encouragement only requires a pat on the back or a gesture now and then. 211

It is necessary to recognize that employees are in a defensive posture throughout their work lives. Hardly anything about their employment is under their control. 211

People know whom to trust. 212

I have watched corporations slide down after building a great headquarters building. 213

Deviation and variation were not considered to be a normal part of life. 214

INDEX

ABOUT THE AUTHOR

Philip B. Crosby is internationally acknowledged as a prime mover in the Quality Revolution. Formerly vice president of ITT and chairman of Philip Crosby Associates and its renowned Quality College, he ranks among the most highly respected and sought-after management consultants, lecturers, and educators. In addition, Crosby is one of the best-selling authors of business books, with such outstanding works to his credit as *Quality Is Free* (1.7 million copies sold), *Quality Without Tears* (400,000 copies sold), *Running Things, The Eternally Successful Organization, Leading, Let's Talk Quality,* and *Philip Crosby's Reflections on Quality,* all published by McGraw-Hill. Crosby is based in Winter Park, Florida, where he is chairman of Career IV.